Graphical Data Analysis with R

Chapman & Hall/CRC
The R Series

Series Editors

John M. Chambers
Department of Statistics
Stanford University
Stanford, California, USA

Torsten Hothorn
Division of Biostatistics
University of Zurich
Switzerland

Duncan Temple Lang
Department of Statistics
University of California, Davis
Davis, California, USA

Hadley Wickham
RStudio
Boston, Massachusetts, USA

Aims and Scope

This book series reflects the recent rapid growth in the development and application of R, the programming language and software environment for statistical computing and graphics. R is now widely used in academic research, education, and industry. It is constantly growing, with new versions of the core software released regularly and more than 6,000 packages available. It is difficult for the documentation to keep pace with the expansion of the software, and this vital book series provides a forum for the publication of books covering many aspects of the development and application of R.

The scope of the series is wide, covering three main threads:
- Applications of R to specific disciplines such as biology, epidemiology, genetics, engineering, finance, and the social sciences.
- Using R for the study of topics of statistical methodology, such as linear and mixed modeling, time series, Bayesian methods, and missing data.
- The development of R, including programming, building packages, and graphics.

The books will appeal to programmers and developers of R software, as well as applied statisticians and data analysts in many fields. The books will feature detailed worked examples and R code fully integrated into the text, ensuring their usefulness to researchers, practitioners and students.

Published Titles

Stated Preference Methods Using R, *Hideo Aizaki, Tomoaki Nakatani, and Kazuo Sato*

Using R for Numerical Analysis in Science and Engineering, *Victor A. Bloomfield*

Event History Analysis with R, *Göran Broström*

Computational Actuarial Science with R, *Arthur Charpentier*

Statistical Computing in C++ and R, *Randall L. Eubank and Ana Kupresanin*

Reproducible Research with R and RStudio, *Christopher Gandrud*

Introduction to Scientific Programming and Simulation Using R, Second Edition, *Owen Jones, Robert Maillardet, and Andrew Robinson*

Nonparametric Statistical Methods Using R, *John Kloke and Joseph McKean*

Displaying Time Series, Spatial, and Space-Time Data with R, *Oscar Perpiñán Lamigueiro*

Programming Graphical User Interfaces with R, *Michael F. Lawrence and John Verzani*

Analyzing Sensory Data with R, *Sébastien Lê and Theirry Worch*

Parallel Computing for Data Science: With Examples in R, C++ and CUDA, *Norman Matloff*

Analyzing Baseball Data with R, *Max Marchi and Jim Albert*

Growth Curve Analysis and Visualization Using R, *Daniel Mirman*

R Graphics, Second Edition, *Paul Murrell*

Data Science in R: A Case Studies Approach to Computational Reasoning and Problem Solving, *Deborah Nolan and Duncan Temple Lang*

Multiple Factor Analysis by Example Using R, *Jérôme Pagès*

Customer and Business Analytics: Applied Data Mining for Business Decision Making Using R, *Daniel S. Putler and Robert E. Krider*

Implementing Reproducible Research, *Victoria Stodden, Friedrich Leisch, and Roger D. Peng*

Graphical Data Analysis with R, *Antony Unwin*

Using R for Introductory Statistics, Second Edition, *John Verzani*

Advanced R, *Hadley Wickham*

Dynamic Documents with R and knitr, *Yihui Xie*

Graphical Data Analysis with R

Antony Unwin
University of Augsburg
Germany

CRC Press
Taylor & Francis Group
Boca Raton   London   New York

CRC Press is an imprint of the
Taylor & Francis Group, an **informa** business

A CHAPMAN & HALL BOOK

CRC Press
Taylor & Francis Group
6000 Broken Sound Parkway NW, Suite 300
Boca Raton, FL 33487-2742

First issued in paperback 2022

© 2015 by Taylor & Francis Group, LLC
CRC Press is an imprint of Taylor & Francis Group, an Informa business

No claim to original U.S. Government works

ISBN 13: 978-1-03-247731-2 (pbk)
ISBN 13: 978-1-4987-1523-2 (hbk)

DOI: 10.1201/9781315370088

Publisher's Note
The publisher has gone to great lengths to ensure the quality of this reprint but points out that some imperfections in the original copies may be apparent.

Visit the Taylor & Francis Web site at
http://www.taylorandfrancis.com

and the CRC Press Web site at
http://www.crcpress.com

Contents

x

Preface

Graphical Data Analysis is useful for data cleaning, exploring data structure, detecting outliers and unusual groups, identifying trends and clusters, spotting local patterns, evaluating modelling output, and presenting results. It is essential for exploratory data analysis and data mining. There are several fine books on graphics using R, such as "ggplot2" [Wickham, 2009], "Lattice" [Sarkar, 2008], and "R Graphics" [Murrell, 2011]). These books concentrate on how you draw graphics in R. This book concentrates on why you draw graphics and which graphics to draw (and uses R to do so).

The target readership includes anyone carrying out data analyses who wants to understand their data using graphics. The book can be used as the primary textbook for a course in Graphical Data Analysis or as an accompanying text for a statistics course. Prerequisites for the book are an interest in data analysis and some basic knowledge of R.

The main aim of the book is to show, using real datasets, what information graphical displays can reveal in data. Seeing graphics in action is the best way to learn Graphical Data Analysis. Gaining experience in interpreting graphics and drawing your own data displays is the most effective way forward.

The graphics shown in the book are a starting point. Sometimes more graphics could have been drawn, and alternative graphics could always have been drawn. Readers may have their own ideas of how best to present certain features of the datasets. Although each graphic reveals information contained in its dataset, it is likely that in every case there is more to be discovered. It is certainly one of the aims of each analysis to find out as much as possible about the data. The graphics are not drawn for their own sake, they are drawn to reveal and convey information.

A central idea underlying this book is that many graphics should be drawn. The aim should not have to be to draw a single graphic that summarises everything that can be said about the data. That is too difficult, if not impossible. The aim is to find a number of graphics, maybe even a large number of them, where each contributes something to the overall picture. Just as many photographs of the same object taken from different angles in different lights make it easier for us to grasp a whole object, datasets should be visualised in many different ways.

The emphasis is on exploring datasets first and on presenting results second. Graphical Data Analysis is about using graphics to find results. One way to think about this is to imagine you are looking at a new package in R and it uses a dataset you are not familiar with for the examples in the help. What does the dataset look like? How would you go about finding out what features it has, and how that might affect the use of the methods in the package? What information can you find graphically in the data that a modelling approach should also find? What graphical displays are there that help you understand the results of other people's models, such as the examples given on the help page? This presupposes an active interest on the part of the reader. Roland Barthes, the French structuralist, referred to readerly texts and writerly texts. In a writerly text the reader takes an active role in the construction of meaning. I hope the readers of this book will take an active role in thinking about what graphics show, what information can be gleaned from them, and why they were chosen.

As every dataset used is available in R or one of its packages, information about them can usually be found on the relevant help page, including which variables of what types are involved and how big the dataset is. Ideally there should be a description of why and how it was collected, with references to original sources. Context is important for interpreting results and you have to know your dataset and its provenance. A well-developed sense of curiosity is very helpful in data analysis.

Graphical Data Analysis is an attractive way of working with data. It encourages you to look at many different aspects and to investigate in many different directions. You can be surprised by what you uncover and even by which graphic turns out to be most effective in revealing information. Your results are easy to show to others and are easy to discuss with others.

For any result found graphically, we should try to check what statistical support there is for it, just as we use graphics to review the results of our statistical modelling. Graphical Data Analysis and more traditional statistical approaches complement each other very well and we should take advantage of this.

Acknowledgements

No book on R should omit thanking Robert Gentleman, Ross Ihaka, and all the many R contributors. They have made analysis of data much easier for the rest of us. Thanks also to Hadley Wickham for all his R packages (sometimes referred to as the Hadleyverse), especially for **ggplot2**, and to Yihui Xie for **knitr**, a major help in keeping this book in order. Particular thanks are due to Bill Venables for his words of wisdom and for R advice and code. If any of the book's code looks elegant, then it must be Bill's, and if it looks clumsy, it is certainly mine.

Dennis Freuer, Urs Freund, Katrin Grimm, Harold Henderson, Ross Ihaka, Kary Myers, Alexander Pilhöfer, Maryann Pirie, Friedrich Pukelsheim, Christina Sanchez, Günther Sawitzki, Rolf Turner, Chris Wild, and Aisen Yang read one or more chapters and made many helpful suggestions for improvement, some of which I have been able to adopt. John Kimmel was an encouraging and efficient publisher, who organised several constructively critical reviewers, including Di Cook, Michael Friendly, and Ramnath Vaidyanathan. I would also like to thank the Statistics Department at the University of Auckland for a stimulating and sociable environment in which to work on this book during my sabbatical.

Finally I would like to thank my family for never asking me when the book would be finished and for many other kindnesses.

Augsburg, December 2014 *Antony Unwin*

1

Setting the Scene

'What is the use of a book', thought Alice, without pictures or conversations?'

Lewis Carroll

1.1   Graphics in action

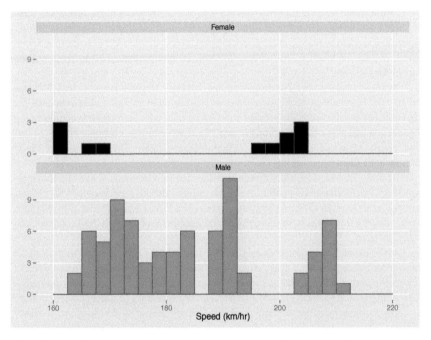

FIGURE 1.1: Histograms of speeds reached at the 2011 World Speed Skiing Championships. Source: www.fis-ski.com. There were more male competitors than females, yet the fastest group of females were almost as fast as the fastest group of males. The female competitors were all either fast or (relatively) slow—or were they?

The code for producing Figure 1.1 is:

```
library(ggplot2); library(ggthemes)
data(SpeedSki, package = "GDAdata")
ggplot(SpeedSki, aes(x=Speed, fill=Sex)) + xlim(160, 220) +
    geom_histogram(binwidth=2.5) + xlab("Speed (km/hr)") +
    facet_wrap(~Sex, ncol=1) + ylab("") +
    theme(legend.position="none")
```

The 2011 World Speed Skiing Championships were held at Verbier in Switzerland. Figure 1.1 shows histograms of the speeds reached by the 12 female and 79 male competitors. As well as emphasising that there were many more competitors in the men's competition than in the women's, the plots show that the fastest person was a man and that a woman was slowest. What is surprising (and more interesting) is that the fastest women were almost as fast as the fastest men and that there were two distinct groups of women, the fast ones and the slow ones. There also appear to be two groups of men, although the gap between them is not so large. All of this information is easy to see in the plots and would not be readily apparent from statistical summaries of the data.

A little more investigation reveals the reason for the groupings: There are actually three different events, Speed One, Speed Downhill, and Speed Downhill Junior. Figure 1.2 shows the histograms of speed by event and gender. We can see that Speed One is the fastest event (competitors have special equipment), that no women took part in the Downhill, and that there was little variation in speed amongst the Juniors.

The reason for the two female groups is now clear: They took part in two different events. The distribution of the men's speeds is affected by the inclusion of speeds for the Downhill event and by the greater numbers of men who competed. It is interesting that there is little variation in speed amongst the 7 women who competed in the Speed One event, compared to that of the 39 men who took part. The women were faster than most of the men.

The code for the plots takes a little getting used to. On the one hand the information would still have been visible with less coding, although perhaps not so clearly. Setting sensible scale limits, specifying meaningful binwidths, and aligning graphics whose distributions you want to compare one above the other with the same size and scales all help.

On the other hand, the plots might have benefitted from more coding to make them look better: adding a title, choosing different colours, or specifying different tick marks and labelling. That is more a matter of taste. This book is about data analysis, primarily exploratory analysis, rather than presentation, so the amount of coding is reduced. Sometimes defaults are removed (like the legends) to reduce unnecessary clutter.

The Speed Skiing example illustrates a number of issues that will recur throughout the book. Graphics are effective ways of summarising and conveying information. You need to think carefully about how to interpret a graphic. Context is important and you often have to gather additional background information. Drawing several graphics is a lot better than just drawing one.

```
ggplot(SpeedSki, aes(Speed, fill=Sex)) +
    geom_histogram(binwidth=2.5)  + xlab("Speed (km/hr)") +
    ylab("") + facet_grid(Sex~Event) +
    theme(legend.position="none")
```

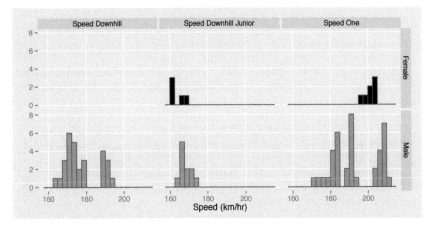

FIGURE 1.2: Histograms of speeds in the 2011 World Speed Skiing Championships by event and gender. There was no Speed Downhill for women. The few women taking part in the fastest event, Speed One, did very well, beating most of the men.

1.2   Introduction

There is no complex theory about graphics. In fact there is not much theory at all, and so the topic is not covered in depth in books or lectures. Once the various graphics forms have been described, the textbooks can pass on to supposedly more difficult topics such as proving the central limit theorem or the asymptotic normality of maximum likelihood estimates.

The evidence of how graphics are used in practice suggests that they need more attention than a cursory introduction backed up by a few examples. If we do not have a theory which can be passed on to others about how to design and interpret informative graphics, then we need to help them develop the necessary skills using a range of instructive examples. It is surprising (and sometimes shocking) how casually graphics may be employed, more as decoration than as information, more for reasons of routine than for reasons of communication.

It is worthwhile, as always, to check what the justly famous John Tukey has to say. In his paper [Tukey, 1993] he summarised what he described as the true purpose of graphic display in four statements:

1. Graphics are for the qualitative/descriptive—conceivably the semiquantitative— never for the carefully quantitative (tables do that better).

2. Graphics are for comparison—comparison of one kind or another—not for access to individual amounts.

3. Graphics are for impact——interocular impact if possible, swinging-finger impact if that is the best one can do, or impact for the unexpected as a minimum——but almost never for something that has to be worked at hard to be perceived.

4. Finally, graphics should report the results of careful data analysis——rather than be an attempt to replace it. (Exploration—to guide data analysis——can make essential interim use of graphics, but unless we are describing the exploration *process* rather than its *results*, the final graphic should build on the data analysis rather than the reverse.)

These are all excellent points, although the last one implies he is emphasising graphics displays for presentation. It is notable that Tukey writes of 'the final graphic' as if there might only be one. In practice he commonly used more than one.

Data analysis is a complex topic. Conclusions are seldom clear-cut, and there are often several alternative competing explanations. Graphics are part of the data analysis process; both the choice of graphics displays and how they are viewed can have an important influence on any conclusions drawn. The subject matter of the study will also affect how any graphics are interpreted. A positively correlated collection of points in a scatterplot may be taken as evidence of a useful association (e.g., income as a function of years of experience) or as evidence of insufficient agreement (e.g., where two methods of measuring the same quantity are compared). In the first case an outlier may be unusual, but nevertheless perfectly plausible, while in the second it may be a clear indication of a faulty measurement.

To appreciate how much can be revealed by even the simplest graphics, it is useful to think of all the different forms a graphic might take and what they might tell you about the data. Consider a barchart showing the frequencies of the three categories of a variable:

- The bars could all be the same height (as you might expect in a scientific study with three groups).

- The bars might have slightly different heights (possibly suggesting some missing values in a scientific study).

- One of the bars might be very small, suggesting that that category is either rare (a particular illness perhaps) or not particularly relevant (support for a minor political party).

- The bars might not follow an anticipated pattern (sales in different regions or the numbers of people with various qualifications applying for a job).

- . . .

There is literally no limit to the number of possibilities once you take into account the different settings the data may have come from. This means that you need to gain experience in looking at graphics to learn to appreciate what they can and cannot show.

As with all statistical investigations it is not only necessary to identify potential conclusions, there has to be enough evidence to support the conclusions. Traditionally this has meant carrying out statistical tests. Unfortunately there are distinct limits to testing. A lot of insights cannot easily be directly tested (Does that outlying cluster of points really form a distinctive group? Is that distribution bimodal?) and even those that can be require restrictive assumptions for the tests to be valid. Additionally there is the issue of multiple testing. None of this should inhibit us from testing when we can, and occasionally a visually tentative result can be shown to have such a convincingly small p-value that no amount of concerns about assumptions can cast much doubt on the result. The interplay of graphics with testing and modelling is effective because the two approaches complement each other so well. The only downside is that while it is usually feasible to find a graphic which tells you something about the results of a test, it is not always possible to find a test which can help you assess a feature you have discovered in a graphic.

1.3   What is Graphical Data Analysis (GDA)?

It is simplest to see what GDA can do by looking at a few examples. These examples are all just initial looks at the datasets to give the flavour of GDA, not complete analyses. In each case we will draw graphical displays of a dataset to reveal some of the information in the data. Graphics are good for showing structure and for communicating results. They are generally easier to interpret than tables (which are good for providing exact values) or statistical reports (which are good for giving estimates and formal comparisons) and convey more qualitative information.

For each example more than one graphic has been drawn. In general it is always better to draw many graphics, offering many different views, to ensure you get as much information out of a dataset as you can. This is part of the open-ended nature of GDA; it is a process, in which you pursue multiple ideas in parallel, just as any investigative process should be. As a result, a graphical data analysis of these examples in practice would involve drawing a lot more graphics, checking to see if there are other features of interest, comparing different versions of various graphics to see which ones work best, and finally settling on a group of graphics to summarise the analysis.

Drawing graphical displays of data is not about selecting the one best graphic, it is about selecting the best set of graphics. In the past drawing graphics required extensive effort and producing one graphic by hand took a lot of time, so it made sense to draw only a few. Nowadays we can draw graphics very quickly, many of the basic options are sensibly chosen by software defaults, and the overall quality is extremely high. It is still a good idea to be selective about which graphics you keep and especially about which graphics you show to others, but the main aim is to uncover information and there is every reason to draw more graphics rather than fewer when doing GDA.

With presentation graphics you prepare one graphic for many potential viewers. You need experience in deciding which graphic to present and expertise in how to draw it well. With GDA you prepare many graphics for one viewer, yourself, and your aim is to uncover the information hidden in the data. You need expertise in choosing a set of informative graphics and experience in interpreting graphics.

The Iris dataset

The dataset contains information on three species of iris. There are 50 plants for each species and measurements in cms for four attributes of each of the 150 plants. Originally the dataset was used by Fisher to illustrate linear discriminant analysis [Fisher, 1936]. (It has been used for all manner of other analyses since.) Figure 1.3 shows a default histogram of petal length, one of the four attributes. There appear to be two clearly different groups in the dataset.

```
ggplot(iris, aes(Petal.Length)) + geom_histogram()
```

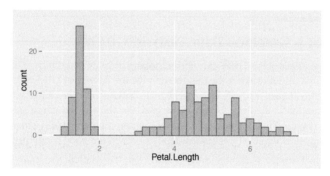

FIGURE 1.3: A histogram of petal lengths from Fisher's iris dataset. The data divide into two distinct groups.

We can look at a plot of the two petal attributes together, petal length and petal width. Figure 1.4 shows that there is a very strong relationship between these two attributes, providing further convincing evidence of at least two distinct groups of flowers. The colouring by species shows that the lower group are all setosa, that the upper group is made up of both versicolor and virginica flowers, and that these two groups are moderately well separated by their petal measurements.

The *iris* dataset is so well known that many readers will be familiar with this information. Imagine, however, that you wanted to present this information to someone who did not already know it. Are there better ways than simple graphics?

```
library(ggthemes)
ggplot(iris, aes(Petal.Length, Petal.Width, color=Species)) +
    geom_point() + theme(legend.position="bottom") +
    scale_colour_colorblind()
```

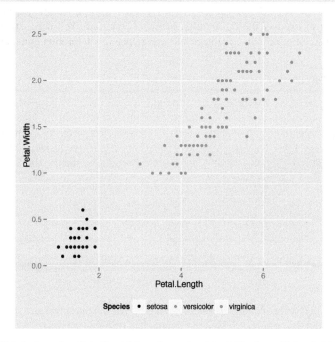

FIGURE 1.4: A scatterplot of petal lengths and petal widths from Fisher's iris dataset with the flowers coloured by species. The two variables are highly correlated and separate setosa clearly from the other two species. The colours used do not reflect the real colours of the species, which are all fairly similar.

Student Admissions at UC Berkeley dataset

The data concern applications to graduate school at Berkeley for the six largest departments in 1973 classified by admission and sex. One of the reasons to study the data was to see whether there was any gender bias in the admission of students. Figure 1.5 shows separate barcharts for the three variables (department, gender, and whether admitted or not).

```
library(gridExtra)
ucba <- as.data.frame(UCBAdmissions)
a <- ggplot(ucba, aes(Dept)) + geom_bar(aes(weight=Freq))
b <- ggplot(ucba, aes(Gender)) + geom_bar(aes(weight=Freq))
c <- ggplot(ucba, aes(Admit)) + geom_bar(aes(weight=Freq))
grid.arrange(a, b, c, nrow=1, widths=c(7,3,3))
```

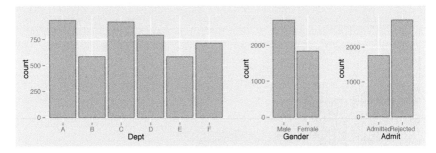

FIGURE 1.5: Numbers of applicants for Berkeley graduate programmes in 1973 for the six biggest departments. The departments had different numbers of applicants. Overall more males applied than females and fewer applicants were admitted than rejected.

It is immediately apparent that different numbers applied to the departments, ranging from about 600 to 900, that there were more male applicants than female ones, and that more applicants were rejected than admitted. None of this is exactly rocket science and all these comments could equally well have been derived from summary statistics (with the additional benefit of knowing that the reported range of departmental applicant numbers was actually from 584 to 933).

Graphics make these points directly and give an overview that is easier to remember than sets of numbers. Graphics are good for qualitative conclusions and often that is what is primarily wanted. Of course, precise numbers may be useful as well, and the two approaches complement one another.

The ordering of the departments is alphabetic, based on the coding used, and other orderings may be more informative. The widths of the plots were controlled to avoid wide bars, a common issue with default charts. The vertical scales might also have been made equal for all three charts (they are close for the last two by chance) to reflect that they display the same cases. Then the differences between the departments would have been downplayed.

The main aim of the study was to examine the acceptance and rejection rates by gender. For the six departments taken together the acceptance rate for females was just over 30% and for males just under 45%, suggesting that there may have been discrimination against females. Results by department are shown in Figure 1.6, where the widths of the bars are proportional to the numbers in the respective groups.. In four of the six departments females had a higher rate of acceptance. This is an example of Simpson's paradox.

```
library(vcd)
ucb <- data.frame(UCBAdmissions)
ucb <- within(ucb, Accept <-
            factor(Admit, levels=c("Rejected", "Admitted")))
doubledecker(xtabs(Freq~ Dept + Gender + Accept, data = ucb),
            gp = gpar(fill = c("grey90", "steelblue")))
```

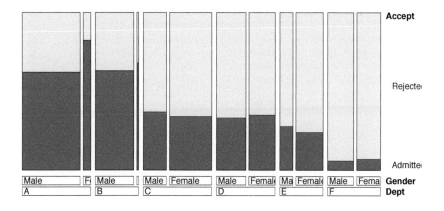

FIGURE 1.6: Acceptance rates at Berkeley by department and gender. Overall males were more likely to be accepted, but four of the six departments were more favourable to females. A doubledecker plot has been used to show the numbers of each gender applying to the departments. Relatively fewer females applied to departments A and B, while relatively fewer males applied to departments C and E.

Pima Indians diabetes dataset

This dataset has been used often in the machine learning literature and it can be found on the web in the excellent UCI library of machine learning datasets [Bache and Lichman, 2013]. The version from R used here, *Pima.tr2* in the **MASS** package, is a training dataset of size 300 with zero values recorded as NA's (a standard missing value code). There are six continuous variables and default histograms are shown for all of them in Figure 1.7.

  The distributions of three variables (plasma glucose, blood pressure, and body mass index) look roughly symmetric. The variable skin thickness has at least one outlier and the diabetes pedigree function distribution is skew, possibly with outliers. The age histogram shows that most women were young with an age distribution like half of a classical age pyramid.

```
data(Pima.tr2, package="MASS")
h1 <- ggplot(Pima.tr2, aes(glu)) + geom_histogram()
h2 <- ggplot(Pima.tr2, aes(bp)) + geom_histogram()
h3 <- ggplot(Pima.tr2, aes(skin)) + geom_histogram()
h4 <- ggplot(Pima.tr2, aes(bmi)) + geom_histogram()
h5 <- ggplot(Pima.tr2, aes(ped)) + geom_histogram()
h6 <- ggplot(Pima.tr2, aes(age)) + geom_histogram()
grid.arrange(h1, h2, h3, h4, h5, h6, nrow=2)
```

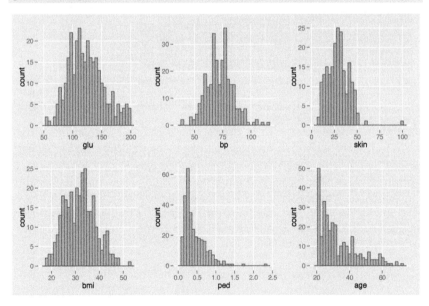

FIGURE 1.7: Histograms of the six continuous variables in *Pima.tr2*. There are a few possible outlying values. Two of the variables have skew distributions.

Rather than drawing a lot of histograms to get an impression of the variables, it takes up less space to draw boxplots. Since R automatically uses the same scale for all boxplots in a window, they have to be standardised in some way first and this can be achieved using the `scale` function. Figure 1.8 shows the result with the outliers coloured in red for emphasis.

```
library(dplyr)
PimaV <- select(Pima.tr2, glu:age)
par(mar=c(3.1, 4.1, 1.1, 2.1))
boxplot(scale(PimaV), pch=16, outcol="red")
```

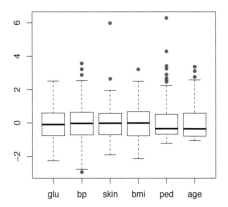

FIGURE 1.8: Scaled boxplots of the six continuous variables in a version of the Pima Indians dataset in R. A couple of big outliers are picked out, as is a low outlier on bp (blood pressure). The distributions of the last two variables, ped (diabetes pedigree) and age, are skewed to the right.

There are several outliers (including a couple of extreme ones) and boxplots are better for showing that than histograms. The last two variables are clearly not symmetric. Two facts should be borne in mind: The boxplots are not to be compared with one another, drawing them all together in one plot is primarily a time and space saving exercise. The scaling just transforms each variable to have a mean of zero with a standard deviation of one and nothing more, so the same points are identified as outliers as would be in the equivalent unscaled plots. This display tells us a little about the shapes of the distributions, but not much, and nothing about the missing values in the data, a potentially important feature. Of course, the histograms told us nothing about the missing values either. Plots for missing values are discussed in §9.2.

The two sets of displays, histograms and boxplots, have given us a lot of information about the variables in the dataset. A scatterplot matrix, as in Figure 1.9, tells us even more.

We can see that only two variables are strongly associated, `bmi` and `skin`, and that that association would be even better were it not for the outlying skin thickness measurement. All of this is valuable information, which helps us to understand the kind of data we are dealing with.

```
library(GGally)
ggpairs(PimaV, diag=list(continuous='density'),
        axisLabels='show')
```

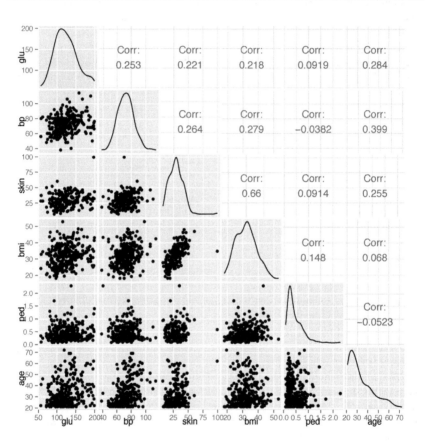

FIGURE 1.9: A scatterplot matrix of the six continuous variables for the same Pima Indians dataset. Only one of the scatterplots shows a strong association (`bmi` and `skin`). The two extreme outlying values, one `skin` and one `ped` measurement, make it harder to see what is going on, as there is less space left for the bulk of the data.

GDA in context

GDA does not stand on its own. As has already been said (and should probably be repeated several times more), any result found graphically should be checked with statistical methods, if at all possible. Graphics are commonly used to check statistical results (residual plots being the classic example) and statistics should be used to check graphical results. You might say that seeing is believing, but testing is convincing.

Graphics are for revealing structure rather than details, for highlighting big differences rather than identifying subtle distinctions. Edwards, Lindman and Savage wrote of the interocular traumatic test; you know what the data mean when the conclusion hits you between the eyes [Edwards et al., 1963]. They were referring to statistical analyses in general, but it is particularly relevant for graphics. Their article continued: "The interocular traumatic test is simple, commands general agreement, and is often applicable; well-conducted experiments often come out that way. But the enthusiast's interocular trauma may be the skeptic's random error. A little arithmetic to verify the extent of the trauma can yield great peace of mind for little cost."

Approximate figures suffice for appreciating structure, there is no need to provide meticulous accuracy. If, however, exact values are needed—and they often are—then tables are more useful. Graphics and tables should not be seen as competitors, they complement one another. With printed reports sometimes difficult choices have to be made about whether to include a graphic or a table. With electronic reports the question becomes how to include both gracefully and effectively.

Graphical Data Analysis is obviously appropriate for observational data where the standard statistical assumptions that are needed for model building may not hold. It can also be valuable for analysing experimental data. There may be patterns by time or other factors that were not expected. In medical studies, the balance of control and treatment groups has to be checked.

This book concentrates on exploratory graphics, using graphics to explore datasets to discover information. Experience gained in looking at and interpreting exploratory graphics will be valuable for looking at all kinds of graphics associated with statistics, including diagnostic graphics (for checking models) and presentation graphics (for displaying results).

The importance of data analysis in general is sometimes underplayed, because there is little formal theory and because the results that are found may appear obvious in retrospect. Effective graphical analysis makes things seem obvious, the effort involved in making the graphical analysis effective is not so obvious. In his poem "Adam's Curse" W. B. Yeats wrote of the amount of work that went into getting a line of poetry right:

Yet if it does not seem a moment's thought,
Our stitching and unstitching has been naught.

That is an appropriate analogy here.

1.4   Using this book, the R code in it, and the book's webpage

A graphic is more than just a picture and every display in the book should convey some information about the dataset it portrays. There should be some description of what you can see accompanying every graphic. In order to ensure that all discussions of graphics are either on the same page as the graphic itself or on the opposite page, gaps have been left on some pages. This is intentional, as however irritating gaps may be, it always seems more irritating having to turn pages backwards and forwards to flip between displays and descriptions.

Like anything else, using graphics effectively is mostly a matter of practice. Study and criticise the examples. Test out the code yourself—it is all available on the book's webpage, rosuda.org/GDA. You can experiment with it while you are reading the book, just copy and paste the code into R. Vary the size and aspect ratio of your graphics, vary the scaling and formatting, vary the colours used. Draw lots of graphics, see what you get, and decide what is most effective for you in making it easy to recognise information.

Work through the exercises at the ends of the chapters. All these exercises use datasets available in R or in one of the packages associated with R. There are no purely technical exercises, they all require consideration of the context involved. The goal is not just to draw graphics successfully, but to interpret the resulting displays and deduce information about the data. Some exercises are more open-ended than others and you should not expect definitive answers to all of them. The best approach is to try several versions of each graphic and to work with sets of graphics of different types, not just with individual ones. Doing the exercises is highly recommended— to become experienced in carrying out Graphical Data Analysis, you need to gain experience in looking at graphics.

There are far too many R packages to load them all. It makes sense to ask readers to load the packages that are used more often in the book instead of repeatedly referring to them in the text. Please ensure you have the following packages loaded:

ggplot2 for graphics based on ideas from "The Grammar of Graphics". Most of the book's graphics use these ideas.

gridExtra for arranging graphics drawn with **ggplot2**.

ggthemes for its colour blind palette.

dplyr for advanced and transparent data manipulation capabilities.

GGally for additional graphics in **ggplot2** form, including parallel coordinate plots.

vcd for a range of graphics for categorical data.

extracat for multivariate categorical data graphics and for missing value patterns.

Some of these packages load further packages via a namespace. To check the state of your R session you can use the function sessionInfo().

Loading packages in advance will mean that the functions in these packages can be immediately used, and that any datasets supplied with the packages are to hand. For datasets in other packages there are two cases to consider. With `LazyData` packages datasets can be accessed without loading the package, as long as you have the package installed. With (most) other installed packages a `data` statement is needed, an R function for making a dataset available. To avoid repetition, datasets are generally only loaded once in each chapter.

Many graphics are improved by an appropriate choice of window size, informative labelling, sensible scaling, good positioning (for instance in multiple graphics displays) and other details, which are more about presentation than the graphic itself and may require much more code than you might expect. This book is about exploratory graphics so the code is mainly restricted to the graphics essentials.

You may find some graphics too small (or too big). If so, redraw them yourself and experiment to find what looks best to you. Space and design restrictions in a printed book can hamper displays. For some of the graphics the code includes adjustments to improve the look of the default versions. Graphics for exploration are usually only on display temporarily, while graphics for presentation, especially in print, are more permanent. Nevertheless a little enhancing helps to avoid occasional displeasing elements in exploratory graphics. It is often a matter of taste and you should develop your own style of graphics for your own use. For presenting to others you need to think of their needs and expectations as well. Some general advice on coding graphics in R is given in Chapter 13.

Code listings for every plot are given in the book and on the book's webpage for downloading. The code is not explained in detail, so if an option choice puzzles you, check the help file for the function, especially the examples there. With R there are always several ways of achieving the same goal and you may find you would have done things differently. The end result is the important thing.

Graphical functions in R can offer very many options and working out what effects they have, especially in combination, can be complicated. It would be nice to be able to say that all functions follow the same rules with similarly defined parameters. Languages are rarely consistent like that, and R is no exception.

No formal statistical analyses are carried out in this text, as there are already many fine books covering statistical modelling. A list of suggested references is given at the end of Chapter 2 and there are a few remarks at the end of each chapter. Readers are encouraged to look for statistical methods to complement their graphical analyses. You should be able to find the tools to do so in R and its many packages. There is often a number of ways offered to carry out particular analyses, each with its own advantages (and possibly disadvantages), so no recommendations are made here.

The final exercises in each chapter are labelled "Intermission" and are intended to be a break and a distraction. Perhaps they would have been better labelled "And now for something completely different". At any rate, I hope they lead you to some interesting visual discoveries and to developing your visual skills in many other directions.

Main points

1. Graphical Data Analysis uses graphics to display and interpret data to reveal the information in a dataset. It is an exploratory tool rather than a confirmatory one.

2. Simple plots can reveal useful information about datasets.
   Figure 1.1 showed a surprising feature of the Speed Skiing Championships and Figure 1.2 explained it. Figure 1.3 showed the two groups of flowers with quite different petal lengths. Figure 1.5 showed that more males than females applied to the Berkeley graduate program. Figure 1.8 showed that there are some extreme outliers in the Pima Indians dataset.

3. Scales and formatting of plots are important.
   Using the same scales for the two histograms in Figure 1.1 and aligning them above one another is essential for conveying the information in the plots effectively. Figure 1.3 is an informative histogram for the iris variable, as we can see that there are two groups and the distributions of lengths within them. Barcharts with different numbers of categories, but from the same dataset, have different default scales and so care is necessary with interpretations across plots (Figure 1.5). Comparing distribution forms for differently scaled variables needs some standardisation first (Figure 1.8).

4. Different plots give different views of the data.
   While Figure 1.3 displays the distribution of petal lengths in the *iris* dataset, Figure 1.4 shows the close relationship between petal length and petal width. Figure 1.7 shows the distribution shapes of the Pima Indian variables and Figure 1.8 emphasises outliers. Figure 1.9 shows that only two of the variables in the dataset appear to be strongly related.

Exercises

More detailed information on the datasets is available on their help pages in R.

1. **Iris**
   How would you describe this histogram of sepal width?

   ```
   ggplot(iris, aes(Sepal.Width)) +
          geom_histogram(binwidth=0.1)
   ```

2. **Pima Indians**
   Summarise what this barchart shows:

   ```
   ggplot(Pima.tr2, aes(type)) + geom_bar()
   ```

3. **Pima Indians**
   Why is the upper left of this plot of numbers of pregnancies against age empty?

   ```
   ggplot(Pima.tr2, aes(age,npreg)) + geom_point()
   ```

4. **Estimating the speed of light**
   There are 100 estimates of the speed of light made by Michelson in 1879, composed of 5 groups of 20 experiments each (dataset *michelson* in the **MASS** package).

   (a) What plot would you draw for showing the distribution of all the values together? What conclusions would you draw?
   (b) What plots might be useful for comparing the estimates from the 5 different experiments? Do the results from the 5 experiments look similar?

5. **Titanic**
   The liner Titanic sank on its maiden voyage in 1912 with great loss of life. The dataset is provided in R as a table. Convert this table into a data frame using `data.frame(Titanic)`.

   (a) What plot would you draw for showing the distribution of all the values together? What conclusions would you draw?
   (b) Draw a graphic to show the number sailing in each class. What order of variable categories did you choose and why? Are you surprised by the different class sizes?
   (c) Draw graphics for the other three categorical variables. How good do you think these data are? Why are there not more detailed data on the ages of those sailing? Even if the age variable information (young and old) was accurate, is this variable likely to be very useful in any modelling?

6. **Swiss**
   The dataset *swiss* contains a standardized fertility measure and various socio-economic indicators for each of 47 French-speaking provinces of Switzerland in about 1888.

   (a) What plot would you draw for showing the distribution of all the values together? What conclusions would you draw?
   (b) Draw graphics for each variable. What can you conclude from the distributions concerning their form and possible outliers?
   (c) Draw a scatterplot of `Fertility` against % `Catholic`. Which kind of areas have the lowest fertility rates?
   (d) What sort of relationship is there between the variables `Education` and `Agriculture`?

7. **Painters**
   The dataset *painters* in package **MASS** contains assessments of 54 classical painters on four characteristics: composition, drawing, colour, and expression. The scores are due to the eighteenth century art critic de Piles.

   (a) What plot would you draw for showing the distribution of all the values together? What conclusions would you draw?
   (b) Draw a display to compare the distributions of the four assessments. Is it necessary to scale the variables first? What information might you lose, if you did? What comments would you make on the distributions individually and as a set?
   (c) What would you expect the association between the scores for drawing and those for colour to be? Draw a scatterplot and discuss what the display shows in relation to your expectations.

8. **Old Faithful**
   The dataset *faithful* contains data on the time between eruptions and the duration of the eruption for the Old Faithful geyser in Yellowstone National Park, USA.

   (a) Draw histograms of the variable `eruptions` using the functions `hist` and `ggplot` (from the package **ggplot2**). Which histogram do you prefer and why? `ggplot` produces a warning, suggesting you choose your own binwidth. What binwidth would you choose to convey all the information you want to convey in a clear manner? Would a boxplot be a good alternative here?

   (b) Draw a scatterplot of the two variables using either `plot` or `ggplot`. How would you summarise the information in the plot?

9. **Intermission**
   Van Dyck's *Charles I, King of England, from Three Angles* belongs to the *Royal Collection* in Windsor Castle. What is gained from having more than one view of the King?

2

Brief Review of the Literature and Background Materials

A picture shows me at a glance what it takes dozens of pages of a book to expound.

Ivan Turgenev (*Fathers and Sons*)

Summary

Chapter 2 reviews some of the available literature on graphics for data analysis and statistics, gives a brief overview of alternative software to R for graphics (as at the time of writing), and explores what relevant material is available on the web: discussions, graphics and datasets. Finally there is a list of texts covering statistical models that might be used in conjunction with the graphical approach described in the book.

2.1 Literature review

Progress in graphics is bedevilled by the lack of theory. Bertin's classic "Semiology of Graphics", originally published in French in 1967, recently reissued by ESRI Press [Bertin, 2010], and Wilkinson's "Grammar of Graphics" [Wilkinson, 2005] are two of the few works that attempt to make substantial contributions. Bertin is not an easy read and is now old-fashioned in many ways, being essentially from a pre-computing age, but the book contains many interesting ideas. It has a strong geographic emphasis. Wilkinson describes a formal structure and includes many attractively drawn displays, although the reasons for drawing the graphics are seldom discussed. Hadley Wickham developed his R package **ggplot2** using the "Grammar of Graphics" and has shown how useful Wilkinson's grammar can be in practice, that it is more than a theoretical construct.

There are far too many books to mention every one that offers good advice on how to draw graphics (let alone to mention all that offer any advice). Tufte's books, especially his first one [Tufte, 2001], are an excellent starting point, containing many attractive and instructive examples as well as some cautionary ones. Tufte offers general principles and discusses both graphics for data analysis and statistics and also what is now called Information Visualisation. Cleveland's advice ([Cleveland, 1993] and [Cleveland, 1994]) is more specific and directed at statistical graphics. Wainer has written a number of enlightening books ([Wainer, 1997], [Wainer, 2004] and [Wainer, 2009]) illustrating with a range of real examples how graphics can be used to reveal information in data. [Robbins, 2005] is worth looking through for practical advice and [Few, 2012], while taking a business-oriented line, supplies forceful opinions in a strongly argued text. For analysts working in the Life Sciences [Krause and McConnell, 2012] offers sound advice and discussion of many real applications.

For graphics in R there is the book [Murrell, 2005], which appeared in an extensively revised second edition in 2011. It covers the full range of R graphics and in particular it explains the details of Murrell's **grid** graphics package, which he implemented as a more structured alternative to the original graphics system in R. The book is fairly technical, but essential for anyone wanting to write graphics using **grid** and valuable for anyone wanting guidance on how graphics in R work. Wickham's **ggplot2** package and Sarkar's **lattice** package are based on **grid**. The additional power and flexibility of **grid** comes at the expense of occasional slowness. Wickham [Wickham, 2009] and Sarkar [Sarkar, 2008] have both written useful books on their packages, although given the continual improvements and changes being made, it is probably best to consult the packages' websites rather than the books.

If you want a specific form of graphic that is not directly available or want to amend or embellish a particular graphic, and these books are not enough, then one of the R cookbooks may help. There are at least two for graphics, [Mittal, 2011] and [Chang, 2012]. In addition, there is the German book [Rahlf, 2014], which provides extensive code examples for producing elegant images. Finally, given that drawing graphics often requires restructuring the data first, it may be helpful to consult one of the many textbooks and resources available for the R language itself.

For graphics not specifically tied to R there have been a number of interesting books in recent years. The visualization volume [Chen et al., 2008] in the Handbook of Computational Statistics series offers a collection of articles from authors with many different views of data visualization. The book "Graphics of Large Datasets" [Unwin et al., 2006] considers the problems of visualising datasets that are a lot bigger than the datasets graphics texts usually discuss. [Inselberg, 2009] treats parallel coordinate plots in detail, emphasising their geometric properties. [Cook and Swayne, 2007] covers dynamic graphics with special reference to the GGobi software for rotating plots. In [Theus and Urbanek, 2007] the authors give an overview of interactive graphics for data analysis including several in-depth case studies. To see how much progress has been made, it is worth looking at "Graphical Exploratory Data Analysis" [DuToit et al., 1986]. Both the ease of producing graphics and the quality of graphics produced have improved enormously in the last twenty-five years.

As this book is about graphics for data analysis and statistics, the emphasis is on the relevant statistical literature. There is additionally an extensive literature on Information Visualisation, which overlaps in interest, although not always in approach, with the books mentioned already. The books [Spence, 2007] and [Bederson and Shneiderman, 2003] are both worth consulting.

There are important factors which affect graphical displays that have nothing to do with data analysis per se. Colour, perception, psychology all play critical roles and effective design should take account of them [Ware, 2008]. There are valuable books on design in general such as [Norman, 1988]. Design is difficult, and much of the theory in these areas is still too far removed from practice to have a direct influence. Theories can explain why something does not work well and they can provide rational support for principles. That does not mean that they can offer specific advice on how to tackle particular tasks. Graphics are still very much a matter of taste.

2.2   Interactive graphics

It would be misleading to talk about GDA without at least some discussion of interactive graphics. Interaction means being able to interact directly with the statistical components of a graphic display. This makes graphical analyses faster and more flexible. The corresponding disadvantages are the difficulty of recording what has been done and the lack of presentation quality reproduction. This will change, and the increasing use of the web for presenting analyses will likely have a big influence.

There are interesting problems to be solved, notably how to formalise interactive graphics, how to provide an intuitive interface for the software, and how to provide results of analyses interactively in a structured form for occasional users. Querying values, zooming in, and reformatting graphics are sensible starting points and much can be achieved with RStudio's Shiny. More is possible, in particular the linking of several associated graphics for the same dataset, which the RCloud project of a group at AT&T provides [Urbanek et al., 2014].

Dynamic graphics, where the displays are animated in some way, for instance rotating through low-dimensional projections of high-dimensional data, is an attractive subset of interactive graphics.

2.3   Other graphics software

Lots of people draw their graphics in Excel, and why not? For many purposes it is a very useful tool. It offers a large variety of alternative graphic forms and considerable numbers of options for amending the appearance of the graphics. Since it is easy to record data in a spreadsheet, Excel is a natural tool for small projects, even though it offers only a limited range of statistical methods. As with all software, it is better for some graphics than others (histograms are difficult to draw in Excel).

The 'big' statistics packages like SAS and SPSS provide extensive graphics capabilities and, like Excel, aim more for preparing presentation graphics than exploratory graphics. They are capable of handling large and complex datasets, provide extensive and substantial statistical capabilities, and can be used to set up regularly operating analysis systems as well as to carry out one-off studies. There are also commercial packages for presentation graphics, some more closely associated with statistical tools than others.

Although R offers interactive graphics resources through the **iplots**, **rggobi**, and **ggvis** packages, it is more for static graphics. There have been a number of commercial interactive graphics softwares, notably Data Desk, JMP, Spotfire, and Tableau. They all have their strengths and weaknesses. Data Desk was very innovative for its time and is still impressive. JMP has the most powerful statistical tools of the group and offers a tightly integrated system. Spotfire arose out of work in Shneiderman's group in Maryland. Tableau is the newest of the group and originated from work at Stanford. On the research side of interactive graphics there are tools like Mondrian and GGobi. All these software packages have been for general applications. Recently there have been some tools designed for particular applications made available on the web, such as Gapminder for displaying scatterplots animated over time or Wordle for word clouds.

The main advantage of concentrating on R is the integrated access to R's extensive range of statistical models and tools.

2.4   Websites

The web changes so quickly that anything written to-day is liable to be out-of-date tomorrow. Nevertheless, several of these websites have been around a while and there are many crosslinks between them. New sites that are any good will doubtless quickly be linked, and once you have found a starting point you should be able to find further interesting sites. Failing that, there is always Google or whatever may take over from them.

A number of websites have sprung up around the theme of data visualization encouraging contributions from readers. Both the *FlowingData* [Yau, 2011] and *Junk Charts* [Fung, 2011] websites discuss visualisations of data and have many interesting examples, often taken from the media. *Statistical Graphics and More* [Theus, 2013] is a similar site. *Many Eyes* [IBM, 2007] lets users create visualisations of datatsets on their site and upload their own datasets for others to visualise. Unsurprisingly the resulting graphics are of a very mixed standard. *Visualizing.org* [Viz, 2011] sees itself as a forum, where readers can discuss visualisation, contribute graphics, and find data. It appears to be more for designers than statisticians. The British site *Improving data visualisation for the public sector* [OCSI, 2009] wants to do just that and includes examples, case studies, and guides.

Tufte includes a discussion page on his website [Tufte, 2013], covering visualisation issues amongst other topics. Gelman's blog *Statistical Modelling, Causal Inference, and Social Science* [Gelman, 2011] often includes debates about graphics and on which alternatives readers prefer. There are sites, which are more Infoviz oriented than statistics oriented, such as *eagereyes* [Kosara, 2011] and *information aesthetics* [Vande Moere, 2011].

The *Gallery of Data Visualisation* [Friendly, 2011] has many classic graphics and helpful supporting historical information. The choice of graphics on display is mildly idiosyncratic, although none the worse for that, and there are enough splendid graphics to suit everyone's taste. The *R Graph Gallery* used to provide a large number of graphics drawn in R and the code used to draw them. Visitors to the site could vote on how good they thought the individual graphics were and it was curious to see which were rated highly and which poorly. The voting probably said more about the voters than about R graphics. The site is no longer maintained. So many websites have sprung up around R that it is impossible to keep track and it would be inadvisable to make firm recommendations given the speed with which things change. All that can be said is that if you are looking for good advice, it is almost certainly available somewhere, just be cautious and check carefully any advice you intend to use.

2.5 Datasets

Many datasets are used in this book, often several times. You need to gain experience in using various kinds of graphics with the same data, and in using graphics for different contexts with different kinds of data. All the datasets are already available in R, in one of its packages, or in the package **GDAdata** accompanying the book. The sources of the datasets and definitions of the variables in them can be found on the corresponding R help pages. To see where a dataset is referred to or analysed by other users of R, you can try one of the search functions on CRAN (`http://cran.r-project.org`).

There are some excellent datasets in R and readers can easily find them to experiment with the data themselves. Unfortunately, there is the disadvantage that the datasets are not always fully documented and that they are sometimes only provided in a cut-down form, for no apparent reason. It is a pity that some R examples are more just for showing how commands work rather than for illustrating how and why you might want to use those commands. More could be done [Unwin et al., 2013]. On the other hand, sorting out the necessary background information and sources for a dataset can involve a lot of hard work (e.g., see the discussions in §3.3 on Galton's and Pearson's family height datasets). Successful data analysis requires sound knowledge of context, so that you can make sense of your results.

A number of datasets are available in several different packages, occasionally under different names and in various versions or formats. For instance, the Titanic dataset used in the book is the one from the **datasets** package. You will also find titanic (**COUNT, prLogistic, msme**), titanic.dat (**exaxtLoglinTest**), titan.Dat (**elrm**), titgrp (**COUNT**), etitanic (**earth**), ptitanic (**rpart.plot**), Lifeboats (**vcd**), TitanicMat (**RelativeRisk**), Titanicp (**vcdExtra**), TitanicSurvival (**effects**), Whitestar (**alr4**), and one package, **plotrix**, includes a manually entered version of the dataset in one of its help examples. The datasets differ on whether the crew is included or not, on the number of cases, on information provided, and on formatting. Versions with the same names in different packages are not identical.

Collecting data requires considerable effort. The individual experiments carried out by Michelson to estimate the speed of light over one hundred years ago are a classic example (cf. Exercise 4 in Chapter 1). There used to be plenty of effort involved in loading datasets into a computer for analysis as well (although far less than needed for gathering the original data). It is amusing to read the instructions for the 1983 DataExpo competition organised by the ASA's Committee on Statistical Graphics: "Because of the Committee's limited (zero) budget for the Exposition, we are forced to provide the data in hardcopy form only (enclosed). (Sorry!) There are 406 observations on the following 8 variables..."

In the early years of the web there were a few sites like Statlib at Carnegie Mellon University or UCI's Machine Learning repository that collected datasets for free use by others. The datasets tended to be small and primarily for teaching purposes (Statlib) or for studying algorithms (UCI). In recent years the situation has totally changed. It is astonishing how much data is now provided on the web. Statistical offices, health departments, election offices, and other official bodies make substantial amounts of data available (for instance, the US Government's site www.data.gov). Sports data are often collected and put up on the web (e.g., the splendid Estonian decathlon webpage www.decathlon2000.com). The British newspaper the *Guardian* runs a *Datastore* [*Guardian*, 2011] where they publish many datasets of public interest ranging from AIDS statistics round the world to more parochial issues like the expenses claimed by British MP's.

Some academic journals require contributors to make their data publicly available. For instance, in 2014 the American Statistical Association's information for authors for its main journal *JASA* included this rule: "Whenever a dataset is used, its source should be fully documented and the data should be made available as an online supplement. Exceptions for reasons of security or confidentiality may be granted by the Editor." [ASA, 2014] We can expect further progress in this direction.

There are R packages to assist you in downloading data from websites that offer their data in particular formats. The book "Data Technologies" [Murrell, 2009] gives advice on gathering and organising web data. When working with web datasets (or indeed with any dataset collected by someone else), it is always a good idea to check the source of the data and to ensure you have sufficient background information. Ideally you should know the aims of the original study from which the data came, how the variables are defined, how the data were collected, and what editing or cleaning of the data has been carried out. Investing time in analysing data that turn out to be flawed is a bad use of your time.

2.6 Statistical texts

Although this book is about graphics for data analysis, there is no claim being made that graphics are enough on their own, far from it. Statistical models are essential for checking ideas found through graphics just as graphics are important for checking results from models.

There are many excellent statistics textbooks that can be recommended for statistical modelling (although perhaps not always for statistical graphics). The following is a personal selection that should cover the models mentioned in this book and more besides.

Sound introductory texts include [Freedman et al., 2007], [De Veaux et al., 2011], and [Maindonald and Braun, 2010]. Other books like [Gelman and Hill, 2006], [Davison, 2008], and [Fahrmeir et al., 2013] assume more knowledge on the part of the reader. All are excellent books. The classic "Modern Applied Statistics with S" [Venables and Ripley, 2002] is still worth reading for its smooth integration of theory, software, and application. Overlaps of statistics with machine learning are well handled by the splendid [Hastie et al., 2001]. Recommended texts covering specific areas of statistics are [van Belle et al., 2004] (Biostatistics), [Tutz, 2012] (Categorical Data Analysis), [Kleiber and Zeileis, 2008] (Econometrics), [Lumley, 2010] (Survey Analysis), [Ord and Fildes, 2013] (Time Series), and, for those wanting a Bayesian approach, [Gelman et al., 2013].

3

Examining Continuous Variables

Normality is a myth; there never was, and never will be, a normal distribution.

Roy Geary ("Testing for Normality")[1]

Summary

Chapter 3 looks at ways of displaying individual continuous variables.

3.1   Introduction

A continuous variable can in theory take any value over its range. (In practice data for continuous variables are generally rounded to some level of measurement accuracy.) Many different plots have been suggested for displaying data distributions of continuous variables and they all have the same aim: to display the important features of the data. Some plots emphasise one feature over another, some are very specialised, some require more deciphering than others. Perhaps the reason there are so many is that there are so many different kinds of features that might be present in the data.

Two possible approaches are to use a range of different plot types or to use a variety of different plots of the same type. In this book the latter approach is mostly favoured and the plots in this chapter are primarily histograms and boxplots. Readers should use what they feel most comfortable with and what works for them. For any plot types you do use, make sure you have plenty of experience in interpreting them. Look at examples of your plots for all kinds of different datasets and get to know both the ways the kinds of features you are interested in are represented and what kinds of structure are emphasised. Textbooks rarely have the space for showing a wide variety of graphics, you have to draw your own.

---

[1]He added "This is an overstatement from the practical point of view..." [Geary, 1947]

28 *Graphical Data Analysis with R*

Figure 3.1 shows a histogram of the percentage share of the vote won by Die Linke in each of the 299 constituencies in the 2009 Bundestag election. Amongst the German political parties there are two on the left of the political spectrum, the SPD, similar to the Labour party in the UK, and Die Linke, literally 'The Left', a party even further to the left. Before drawing the histogram, a new variable is calculated to give the percentage support in each constituency. The binwidth is chosen to be 1 to give bins that are easy to interpret. Thus the height of each bin shows the number of constituencies with a particular percentage of Die Linke supporters. Although they are a party on the left, they are usually given a purple colour in displays of results, as red is used for the SPD.

The party had little support in most of the country, but managed over 10% in a fifth of the constituencies. The areas where they did well turned out to be in the new *Bundesländer* (the old East Germany), in East Berlin, although not in West Berlin, and in the small *Bundesland* called *Saarland*, where their party leader in that election was based, Oscar Lafontaine. You might have guessed at a pronounced regional distribution given the form of the histogram, but not that the country would be so precisely divided up.

```
data(btw2009, package = "flexclust")
btw2009 <- within(btw2009, Linke2 <- 100*LINKE2/valid2)
ggplot(btw2009, aes(Linke2)) + geom_bar(binwidth = 1,
       fill = "mediumpurple") + ylab("") +
       xlab("Percentage voter support for Die Linke in 2009")
```

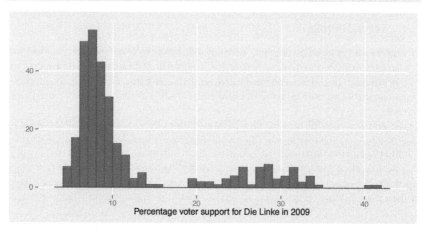

FIGURE 3.1: Percentage support by constituency for Die Linke party in the German election of 2009. The graphic suggests a division of the constituencies into two distinct groups. Further investigation revealed that the constituencies where Die Linke did poorly are all in the old Western Germany, including West Berlin, but excluding Saarland, where the party leader was based.

3.2 What features might continuous variables have?

There might be

Asymmetry the distribution is skewed to the left or right, for instance distributions of income.

Outliers there are one or more values that are far from the rest of the data.

Multimodality the distribution has more than one peak, e.g., both variables in the Old Faithful geyser data.

Gaps there are ranges of values within the data where no cases are recorded. This can happen with exam marks, where there are no cases just below the pass mark.

Heaping some values occur unexpectedly often. Birthweights are a good example [Clemons and Pagano, 1999]. Perhaps there are more important things to do than to weigh the newborn baby to the nearest gram or ounce …

Rounding only certain values (perhaps integers) are found, e.g., age distributions.

Impossibilities values outside the feasible range, for instance negative ages. There is a version of the Pima Indians dataset discussed in 1.3 in the UCI machine learning repository [Bache and Lichman, 2013] in which there are blood pressure and skin thickness measurements of zero.

Errors values that look wrong for one reason or another. In a German car insurance dataset there were drivers whose year of birth gave them ages of less than 16, so they could not have a licence. It was suggested that this might be possible, as insuring someone who never drove would enable them to build up a history of several years of no claims!

Graphics are good for displaying the features that make up the shapes of data distributions. They can provide more and different kinds of information than a set of summary statistics. Obviously it is best to use both approaches.

With a single variable the mean is usually the most important statistic and perhaps no statistical test is used as often as the t-test for testing means. A t-test can be used if the underlying data are from a normal distribution. For small datasets (and the t-test is intended specifically for small samples) data from a normal distribution can look very non-normal, which is why tests of normality have low power and provide little support for t-tests. Fortunately the t-test is fairly robust against non-normality. This should not prevent anyone from at least checking whether the data have some seriously non-normal feature before carrying out analyses. That can best be done graphically.

3.3   Looking for features

This section discusses a number of different datasets, mainly using histograms, to see which features might be present and how they can be found. Whichever display you favour, it is important to study lots of examples to see the various forms of the graphic that can arise and to get experience in interpreting them.

Galton's heights

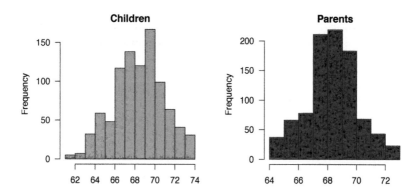

FIGURE 3.2: Histograms of child and midparent heights with default binwidths. The plots have been drawn separately, so the scales are different. Both distributions are roughly symmetric and there is more spread amongst the children. The binwidth has been set to 1 inch. If the data were in cms, then another binwidth would be more suitable, perhaps 2.5 cms.

```
data(galton, package="UsingR")
ht <- "height (in)"
par(mfrow=c(1,2), las=1, mar=c(3.1, 4.1, 1.1, 2.1))
with(galton, {
    hist(child, xlab=ht, main="Children", col="green")
    hist(parent, xlab=ht, main="Parents", col="blue")})
```

Galton famously developed his ideas on correlation and regression using data which included the heights of parents and children. The dataset *galton* in the package **UsingR** includes data on heights for 928 children and 205 'midparents'. Each parent height is an average of the father's height and 1.08 times the mother's height. The daughters' heights are also said to have been multiplied by 1.08. Note that there is one midparent height for each height of a child in this dataset, so that many midparent heights are repeated. (For anyone interested in investigating a detailed dataset with full family information, including the sex and order of the children, see

[Hanley, 2004] and that author's webpage.) Figure 3.2 shows default histograms of the two variables.

Both distributions are vaguely symmetric and there appear to be no outliers. It is difficult to compare the histograms as the scales in the two plots are different. Although the histograms appear to have different binwidths they are actually the same, thanks to R looking for sensible breaks, in this case integers, and only drawing bins for the range of the data. To look for gaps or heaping, it can be useful to draw histograms with very thin bins. Barcharts would also work, although not for large datasets with many individual values. Dotplots are good for gaps, although not for heaping.

From Figure 3.3, histograms with binwidths of 0.1, we can see that only a limited number of values are used for height (and if you go to the source of this version of the dataset given in the R help you find the data were taken from a table, so individual values are not provided). In both histograms there appear to be narrower gaps between values at the boundaries. Drawing up tables of the 'raw' data confirms this and reveals the curiosity that the parent values almost all end in .5 while the child values almost all end in .2. It would be better to use Hanley's version of the dataset. The function `truehist` was used for Figure 3.3, as the binwidth can be set directly. (`hist` uses the number of cells.) As it regards histograms as density estimates, the *y*-axis scale no longer shows the frequencies.

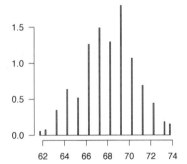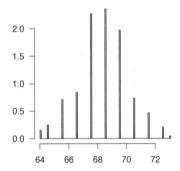

FIGURE 3.3: Histograms of child and midparent heights drawn with a narrow binwidth to demonstrate how few height values actually occur. Clearly the data were aggregated or reported to a limited level of precision. The plots have been drawn separately, so the scales are different.

```
par(mfrow=c(1,2), las=1, mar=c(3.1, 4.1, 1.1, 2.1))
with(galton, {
    MASS::truehist(child, h=0.1)
    MASS::truehist(parent, h=0.1)})
```

To actually compare the distributions a number of plots could be drawn, for instance parallel boxplots. Keeping with the histogram theme, Figure 3.4 displays the histograms one above the other with equal scales and binwidths. The x-axis scale limits were chosen to include the full range of the data and the y-axis limits were chosen by inspection.

In interpreting the data it should be borne in mind that, as Hanley points out, Galton obtained the data "through the offer of prizes" for the "best Extracts from their own Family Records" [Hanley, 2004], so the sample is hardly a random one; that the data have been rounded for tabulation; and that family sizes vary from 1 to 15, so that one midparent data point occurs 15 times in the dataset and 33 points occur only once. The medians have been marked by red vertical lines. Although the means are very similar, the parents' distribution is clearly less variable. This is because each parent value is an average of two values. Interestingly (and perhaps even suspiciously) the standard deviation of the children, 2.52, is almost exactly $\sqrt{2}$ times the standard deviation of the parents, 1.79.

You might think that if the individual parent values were available and the genders of the children were known, as is the case with the full Galton data set available on Hanley's webpage, then the height data distributions would be neatly bimodal with one peak for females and one for males. They are not. Apparently height distributions are rarely like that, as is discussed in [Schilling et al., 2002], which includes some nice photographic examples of putative bimodal height distributions constructed by getting people to stand in lines according to their heights.

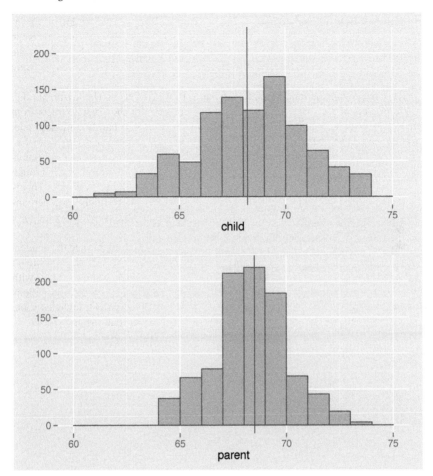

FIGURE 3.4: Histograms of child and midparent heights drawn to the same scale and with medians marked in red. Parents' heights are spread less and their median is slightly higher.

```
c1 <- ggplot(galton, aes(child)) + geom_bar( binwidth=1) +
         xlim(60, 75) + ylim(0, 225) + ylab("") +
         geom_vline(xintercept=median(galton$child),
         col="red")
p1 <- ggplot(galton, aes(parent)) + geom_bar( binwidth=1) +
         xlim(60, 75) + ylim(0, 225) + ylab("") +
         geom_vline(xintercept=median(galton$parent),
         col="red")
grid.arrange(c1, p1)
```

Some more heights—Pearson

There is another dataset of father and son heights in the **UsingR** package. This stems from Karl Pearson and includes 1078 paired heights. His paper with Alice Lee [Pearson and Lee, 1903] includes a detailed description of how the data were collected. Families were invited to provide their measurements to the nearest quarter of an inch with the note that "the Professor trusts to the *bona fides* of each recorder to send only correct results." The dataset was used in the well-known text [Freedman et al., 2007] and a scatterplot of the data is on the book's cover. As the data are given to five decimal places and as each one of the father's heights is a unique value (even though we know there must have been repeats), someone must have jittered the data.

The father and son height distributions can be displayed as histograms with overlaid density estimates (Figure 3.5). Densities can only be successfully overlaid when the histogram scales are densities instead of frequencies, which is why y=..density.. is needed. Both distributions look fairly normal, as we might expect, and, if anything, an initial look would suggest that the heights of the sons look more normal than the heights of the fathers.

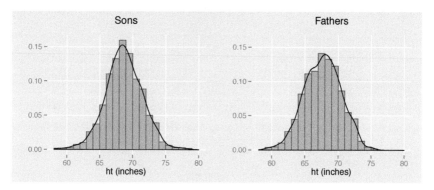

FIGURE 3.5: Histograms and overlaid density estimates of the heights of fathers and sons from the *father.son* dataset in the **UsingR** package. Both distributions look fairly normal. For comparison purposes the plots have been common-scaled.

```
data(father.son, package="UsingR")
c2 <- ggplot(father.son, aes(sheight)) +
         geom_histogram(aes(y = ..density..), binwidth=1) +
         geom_density() + xlim(58, 80) + ylim(0, 0.16) +
         xlab("ht (inches)") + ylab("") + ggtitle("Sons")
p2 <- ggplot(father.son, aes(fheight)) +
         geom_histogram(aes(y = ..density..), binwidth=1) +
         geom_density() + xlim(58, 80) + ylim(0, 0.16) +
         xlab("ht (inches)") + ylab("") +
         ggtitle("Fathers")
grid.arrange(c2, p2, nrow = 1)
```

You have to be careful about jumping to conclusions here, or as Pearson put it in his 1903 paper with Lee: "It is almost in vain that one enters a protest against the mere graphical representation of goodness of fit, now that we have an exact measure of it." Whether we would regard Pearson's test as an "exact measure" nowadays is neither here nor there. For the time, he was justified in his polemically demanding both graphical and analytic approaches. Since those early days, analytic approaches have often been too dominant, both are needed.

If you really want to look at normality, then Q-Q plots are best (Figure 3.6). These suggest that the fathers' heights are more normal than the sons'! Comparing these graphical impressions with some tests (`shapiro.test` for Shapiro-Wilk and the further five normality tests in **nortest**) gives a range of p-values for the fathers from 0.42 to 0.20 (so none are significant) and for the sons from 0.07 to 0.01 (where four of the six tests are significant).

```
par(mfrow=c(1,2), las=1, mar=c(3.1, 4.1, 1.1, 2.1))
with(father.son, {
        qqnorm(sheight, main="Sons", xlab="",
               ylab="", pch=16, ylim=c(55,80))
        qqline(sheight)
        qqnorm(fheight, main="Fathers", xlab="",
               ylab="", pch=16, ylim=c(55,80))
        qqline(fheight) })
```

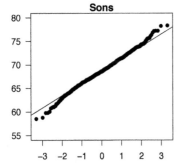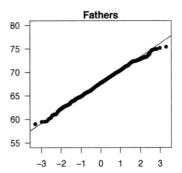

FIGURE 3.6: Q-Q plots of the heights of fathers and sons from the dataset *father.son* with lines added going through the 25% and 75% quantiles. The distribution of the fathers' heights now looks more normal than the distribution of the sons' heights, because of the deviations in the upper and lower tails for the sons.

It is curious that these two famous datasets on heights of parents and children, *galton* and *father.son*, have been altered in opposite ways: Some heights in *galton* have been rounded and some heights in *father.son* have been jittered. It is always better to make the raw data available, even if they need to be amended, 'just in case'.

Scottish hill races (best times)

The *hills* dataset in the package **MASS** contains record times from 1984 for 35 Scottish hill races. The total height gained and the race distance are also included. The dataset is well known, as it is used in the introductory chapter of [Venables and Ripley, 2002]. The **DAAG** package includes some later hills data with more races in various datasets. Figure 3.7 shows a boxplot of the record times for the races in the original dataset. Four races required much longer times than the others and the distribution is skewed to the right (look at the position of the median in the box). If you look at three histograms produced by some default histograms, you will find that the `truehist` default is poor, the `hist` default is a little better, although it still does not pick out the outliers clearly, while the `qplot` histogram (a so-called quick option in **ggplot2**) does direct your attention to the outliers, while at the same time being too detailed for the races with faster record times. Formulating default settings for histograms is hard and it is simpler and more informative to draw a range of histograms with different settings.

```
par(mfrow=c(1,1), mar=c(3.1, 4.1, 1.1, 2.1))
with(MASS::hills,
    boxplot(time, horizontal=TRUE, pch=16, ylim=c(0, 220)))
```

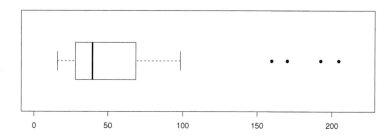

FIGURE 3.7: A boxplot of the record times for the *hills* dataset. The distribution is skew to the right with a few high outliers, races which must have been more demanding than the others.

How are the variables in the Boston dataset distributed?

The Boston housing data is a dataset from 1978 that has been analysed many times. There are two versions in R, the original one in **MASS** and a corrected one in **DAAG**, in which 5 median house values have been 'corrected'. There are 14 variables for 506 areas in and around Boston. The main interest lies in the median values of owner-occupied homes by area and here we will use the original dataset. Default histograms (drawn with either `hist(medv)` or `truehist(medv)`) hint that there might be some interesting structure and that other binwidths or graphics might be useful, while Figure 3.8, drawn with `ggplot`, highlights two features: there are surprisingly many

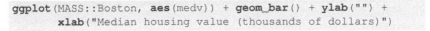

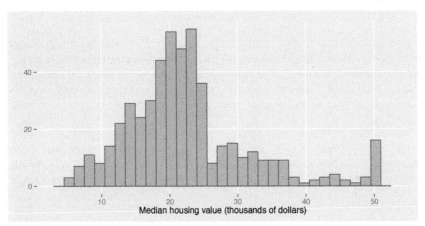

FIGURE 3.8: A histogram of the median values of owner-occupied homes in the *Boston* dataset. The large frequency of the final bin is unusual and there appears to be a sharp decline in the middle.

areas in the final bin and there is a sudden drop in the counts round about the middle. A binwidth matching the data units, say 1 or 2, would be better.

At this stage you really need to examine the data directly, either with a table or with a histogram with a narrow binwidth. A table (`table(medv)`) is not so helpful in this case, but at least it tells us that the data are reported to one decimal place (in thousands of dollars). A histogram with a binwidth of 1 (say with `truehist(medv, h=1)`) confirms that there are very many values in the last bin at 50, but does not tell us much more about the break around 25.

In a full analysis it would now be a good idea to look at the rest of the dataset to see if features observed in the median values variable are related to features in other variables. That will be done in a later chapter, but in the meantime consider Figure 3.9, which displays histograms of all the variables. You can readily see the variety of possible forms a histogram may take and thinking about what each might mean is the subject of Exercise 2 at the end of the chapter. The binary variable `chas` has been included, not because histograms are good for binary variables, but because it is coded numerically. It is easier to leave it in, and the histogram does show the information that only a few areas have a river boundary.

The code needs some explanation. The `melt` function creates a new dataset with all the data in one variable called `BostonValues` and a second variable called `BostonVars` defining which of the original variables a value comes from. The second line of code draws a set of histograms, one for each of the original variables, using facetting, where the option `scales="free"` ensures that each display is individually scaled.

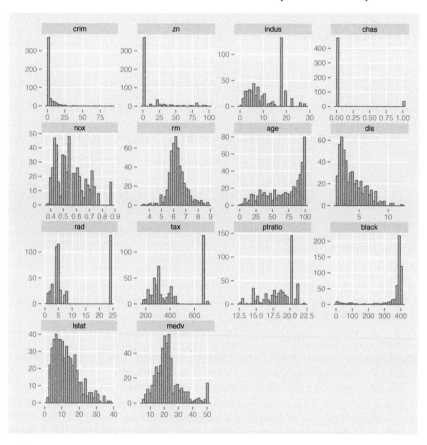

FIGURE 3.9: Histograms of 14 variables from the Boston housing dataset. There are several different histogram forms, each telling a separate story. Default binwidths, dividing each variable's range by 30, have been used. Other scalings could reveal more information and would be more interpretable.

```
library(tidyr)
B2 <- gather(MASS::Boston, BosVars, BosValues, crim:medv)
ggplot(B2, aes(BosValues)) + geom_histogram() + xlab("") +
      ylab("") + facet_wrap(~ BosVars, scales = "free")
```

Plots like these are not ideal. In particular, the default scaling of 30 bins works better for some than for others. Nevertheless, this is a quick and easy way to get an overview of the data and you can always redraw any individual plots which you think deserve a closer look. Note that the vertical scales vary from maxima of 40 to over 400. If you plot the histograms individually, choosing binwidths and scale limits are the main decisions to be taken. Occasionally the anchorpoint, where the first bin starts, and whether the bins are open to the left or to the right can make a difference. The latter becomes an issue if many points lie on bin boundaries. Compare the two histograms produced by default by `hist` (open to the left) and `truehist` (open to the right) for the variable `ptratio`.

```
with(Boston, hist(ptratio))
with(Boston, truehist(ptratio))
```

Histograms of datasets with rounded values need to be checked for these effects. And in case you are wondering, `ggplot` is open to the left like `hist`.

It is worth considering what other plots of the variable `medv` might show. Here are some you could look at:

- Boxplots

```
boxplot(Boston$medv, pch=16)
```

- Jitterered dotplots

```
stripchart(Boston$medv, method="jitter", pch=16)
```

- Stem and leaf plots

```
stem(Boston$medv)
```

- Average shifted histograms

```
library(ash)
plot(ash1(bin1(Boston$medv, nbin=50)), type="l")
```

- Density estimates with a rugplot

```
d1 <- density(Boston$medv)
plot(d1, ylim=c(0,0.08))
rug(Boston$medv)
lines(density(Boston$medv, d1$bw/2), col="green")
lines(density(Boston$medv, d1$bw/5), col="blue")
```

Some of these work much better than others in revealing features in the data. There is no optimal answer for how you find out information, it is only important that you find it. Note that what density estimates show depends greatly on the bandwidth used, just as what histograms show depends on the binwidth used, although histograms also depend on their anchorpoint. One graphic may work best for you and another may work best for someone else. Be prepared to use several alternatives.

Hidalgo stamps thickness

The dataset *Hidalgo1872* is available in the package **MMST**, which accompanies the textbook "Modern Multivariate Statistical Techniques" [Izenman, 2008]. The dataset was first discussed in detail in 1988 in [Izenman and Sommer, 1988]. A keen stamp collector called Walton von Winkle had bought several collections of Mexican stamps from 1872-1874 and measured the thickness of all of them. The thickness of paper used apparently affects the value of the stamps to collectors, and Izenman's interest was in looking at the dataset as a mixture of distributions. There are 485 stamps in the dataset and for some purposes the stamps may be divided into two groups (the years 1872 and 1873-4). We shall examine the full set here first.

The aim is to investigate what paper thicknesses may have been used, assuming that each shows some kind of variability. Figure 3.10 displays a histogram with a binwidth of 0.001 (the measurements were recorded to a thousandth of a millimeter) and two density estimates, one using the bandwidth selected by the density function and one using the direct plug-in bandwidth, dpik, described in [Wand and Jones, 1995], used in the kde function of package **ks**.

The histogram suggests there could be many modes, seven or even eight, depending on how you interpret the three peaks to the left of the distribution. The density estimate from density suggests there are two, and the estimate from kde suggests perhaps seven. In Izenman and Sommer's 1988 paper they found seven modes with a nonparametric approach and three with a parametric one. This is a complex issue and it underlines the value of looking at more than one display.

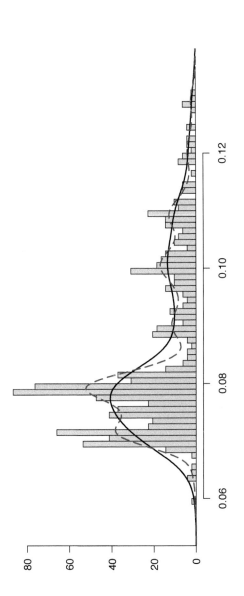

FIGURE 3.10: A histogram and two density estimates of stamp thicknesses from the *Hidalgo1872* dataset. Both the histogram and the red density estimate suggest there are at least 5 modes, implying different production runs or locations.

```
library(KernSmooth)
data(Hidalgo1872, package="MMST")
par(las=1, mar=c(3.1, 4.1, 1.1, 2.1))
with(Hidalgo1872, {
    hist(thickness,breaks=seq(0.055,0.135,0.001), freq=FALSE, main="", col="bisque2", ylab="")
    lines(density(thickness), lwd=2)
    ks1 <- bkde(thickness, bandwidth=dpik(thickness))
    lines(ks1, col="red", lty=5, lwd=2)})
```

How long is a movie?

Most of the datasets in R are not very big. Examples provided by package authors are more for illustration and for showing how methods work than for carrying out full scale data analyses of large datasets. The effort involved in preparing and making available a large dataset in a usable form should never be underestimated.

Nevertheless there are some largish datasets to be found, for instance the dataset *movies* in **ggplot2** with 58,788 cases and 24 variables (which on no account should be confused with the small dataset of the same name in **UsingR** with 25 cases and 4 variables, nor indeed with the larger more up-to-date version with 130,456 films in the package **bigvis**). Strangely, rather modestly, the help page says there are only 28,819 cases. One of the variables is the movie length in minutes and it is interesting to look at this variable in some detail, starting with the default histogram (Figure 3.11) and boxplot (Figure 3.12) using **ggplot2**.

```
ggplot(movies, aes(length)) + geom_bar() + ylab("")
```

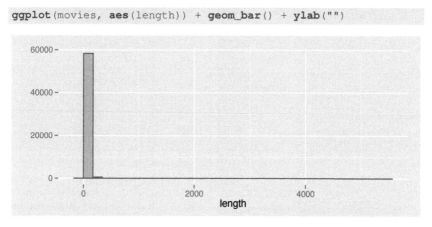

FIGURE 3.11: A histogram of film length from the *movies* dataset. There must be at least one extreme outlier to the right distorting the scale.

The histogram is hopeless, except that it does imply there must be at least one very high value to the extreme right, even if the resolution of the plot is not good enough to see it (with over 58,000 cases in the single bar on the left, one case is never going to be visible on its own, a typical problem for large datasets).

The boxplot does what boxplots do best, telling us something more about the outliers. It appears there are two particularly gross ones, one implying a film length of over three and a half days and the other with a length of about two days. Although you would think these would have to be errors, it is always best to check, if possible. With the help of something like

```
sl <- filter(movies, length > 2000)
print(sl[, c("length", "title")], row.names=FALSE)
```

```
ggplot(movies, aes("var", length)) + geom_boxplot() +
      xlab("")  + scale_x_discrete(breaks=NULL) + coord_flip()
```

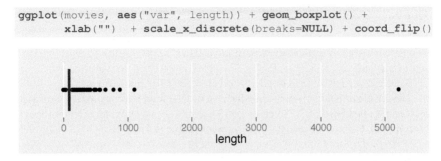

FIGURE 3.12: A boxplot of film length from the *movies* dataset. (An artificial *x* aesthetic, here `"var"`, is needed when drawing single boxplots with **ggplot2**.) There are two very long films, longer than two days, and many that are much longer than the average film.

you can find out the titles of the films and their length and then check if they really exist by Googling them. It turns out they do! Slightly bizarrely, the second longest film in the dataset is called "The Longest Most Meaningless Movie in the World". It must have been a shock for the makers to discover that their film wasn't. Incidentally, this version of the dataset is no longer up-to-date and there are now some even longer films.

Clearly the extreme outliers should be ignored, and for exploring the main distribution of movie lengths it makes sense to set some kind of upper limit. Over 99% are less than three hours in length, so we will restrict ourselves to them. Having got rid of the outliers, another boxplot does not make a lot of sense and a stem and leaf plot or a dotplot (whether jittered or not) is clearly out of the question. A density estimate could be interesting, but might miss possible heaping or gaps, if these features are present. Figure 3.13 shows a histogram with the natural binwidth of 1 minute. (Any other value might miss some feature of the raw data and with such a large dataset there can be no concerns about having too many bins, you just have to ensure that the plot window is wide enough.) Several features now stand out:

1. There are few long films (i.e., over two and a half hours).

2. There is a distinct group of short films with a pronounced peak at a length of 7 minutes.

3. There is, unsurprisingly, a sharp peak at a length of 90 minutes. Perhaps we would have expected an even sharper peak.

4. There is clear evidence of round numbers being favoured; as well as a length of 90 minutes, you also find 80, 85, 95, 100, and so on.

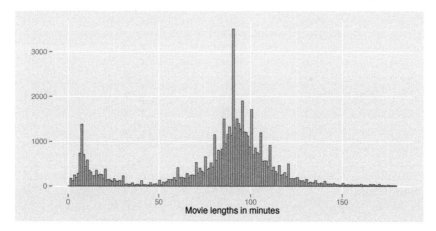

FIGURE 3.13: A histogram of film lengths up to 3 hours from the *movies* dataset.
The most frequent value is 90 minutes and values are often rounded to the nearest 5
minutes. The most frequent time for short films is 7 minutes.

```
ggplot(movies, aes(x = length)) + xlim(0,180) +
    geom_histogram(binwidth=1) +
    xlab("Movie lengths in minutes") + ylab("")
```

3.4   Comparing distributions by subgroups

When a population is made up of several different groups, it is informative to compare the distribution of a variable across the groups. Boxplots are nearly always best
for this, as they make such an efficient use of the space available. Figure 3.1 showed
the voting support of Die Linke party in Germany and claimed that the form was due
to regional differences. Figure 3.14 shows boxplots of the support by Bundesland and
you can see that apart from Berlin, which divides into East and West, the boxplots
split into two distinct groups, basically the new Bundesländer plus Saarland (higher
support) and the old Bundesländer minus Saarland (lower support). Although the
boxplots convey a lot of information in a limited space, they have two disadvantages:
If a distribution is not unimodal (e.g., Berlin), you can't see that, and it is difficult to
convey information on how big the different groups are. Neither is serious, but you
need to be aware of them. The option varwidth sets the boxplot widths proportional to the square root of the number of cases. Another alternative is to draw an
accompanying barchart for the numbers in each group (cf. Figure 4.1).

Figure 3.14 shows a range of possible boxplot shapes. There are outliers above
and below, whiskers of different lengths, and asymmetries of various kinds.

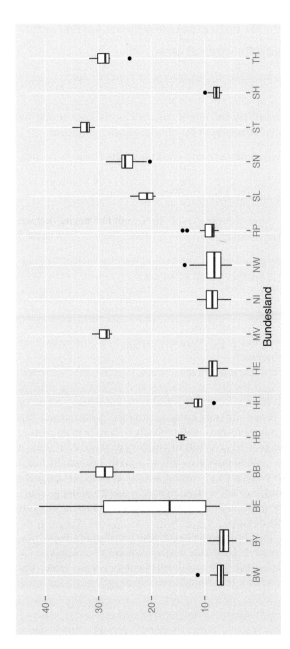

FIGURE 3.14: Boxplots of Die Linke support by *Bundesland*. Standard abbreviations for the *Bundesländer* have been added to the dataset to make the labels readable. In the old East (BB, MV, SN, ST, TH) Die Linke were relatively strong, in the old West they were weak, apart from in Saarland (SL). Berlin (BE), made up of East and West, straddles both groups. Boxplot widths are a function of *Bundesland* size.

```
btw2009 <- within(btw2009, Bundesland <-state)
btw2009 <- within(btw2009, levels(Bundesland) <- c("BW", "BY", "BE", "BB",
   "HB", "HH", "HE", "MV", "NI", "NW", "RP", "SL", "SN", "ST", "SH", "TH"))
ggplot(btw2009, aes(Bundesland, Linke2)) + geom_boxplot(varwidth=TRUE) + ylab(""))
```

3.5   What plots are there for individual continuous variables?

To display continuous data graphically you could use a

histogram grouping data into intervals, and drawing a bar for each interval, shows the empirical distribution.

boxplot displaying individual outliers and robust statistics for the data, useful for identifying outliers and for comparisons of distributions across subgroups.

dotplot plotting each point individually as a dot, good for spotting gaps in the data.

rugplot plotting each point individually as a line, often used as an additional plot along the horizontal axis of another display.

density estimate plotting an estimated density of the variable's distribution, so more like a model than a data display.

distribution estimate showing the estimated distribution function, useful for comparing distributions, if one is always 'ahead' of another.

Q-Q plot comparing the distribution to a theoretical one (most often a normal distribution).

And there are other possibilities too (e.g., frequency polygon, P-P plot, average shifted histogram, shorth plot, beanplot).

R's default for `plot` is to draw a scatterplot of the variable against the case index. This can be useful (e.g., showing if the data have been sorted in increasing order or that the first few values or last few values are different from the others), mostly it is not. Different analysts may favour different kinds of displays, for instance I like histograms and boxplots. Pronounced features will probably be visible in all plots.

For more subtle effects the best approach in exploratory analysis is to draw a variety of plots. There is some general advice to follow, such as histograms being poor for small datasets, dotplots being poor for large datasets and boxplots obscuring multimodality, but it is surprising how often even apparently inappropriate graphics can still reveal information. The most important advice remains—which is why it is now repeated—to draw a variety of plots.

If data are highly skewed it may be sensible to consider transforming them, perhaps using a Box-Cox transformation. Graphical displays can help you appraise the effectiveness of any transformation, but they cannot tell you if they make sense. You should consider the interpretation of a transformed variable as well as its statistical properties.

3.6 Plot options

- *Binwidths (and anchorpoints) for histograms*
  There is an intriguing and impressive literature on the data-driven choice of binwidths for histograms. [Scott, 1992] and articles by Wand (for example, [Wand, 1997]) are reliable sources. In practice there are often good reasons for choosing a particular binwidth that is not optimal in a mathematical sense. The data may be ages in years, or times in minutes, or distances in miles. Using a non-integer binwidth may be mathematically satisfying, but can conceal useful empirical information. It is important to remember that histograms are for presenting data; they are poor density estimators. There are far better approaches for estimating a possible density generating the data. And it is worth bearing in mind that methods for determining optimal histogram binwidth assume a given anchorpoint, i.e., the starting point of the first bin. Both display parameters should really be used for optimisation. In his package **ggplot2** Wickham does not attempt to find any 'optimal' choice, but uses 30 bins and prints a message explaining how to change the binwidth. That is a practical solution.

- *Unequal binwidths*
  When they introduce histograms, some authors like to point out the possibility of using unequal binwidths. While the idea is theoretically attractive, it is awkward to apply in practice and the displays are confusing to interpret. If you still want to do it, consider a variable transformation instead.

- *Bandwidth for density estimates*
  Binwidth is crucial for histograms and bandwidth is crucial for density estimates. There are many R packages offering different bandwidth formulae and it is not obvious which to recommend. It is more effective to experiment with a range of bandwidths. Since you can overlay several density estimates on a single plot, it is easy to compare them, just use different colours to make them stand out.

- *Boxplots*
  Tukey's definition of a boxplot distinguishes between outliers (over 1.5 times the box length away from the box) and extreme outliers (over 3 times the box length away from the box), while many boxplot displays do not. In fact, there are frustratingly many different boxplot definitions around, so you should always confirm which one is used in any plot. Some do not mark outliers, some use standard deviations instead of robust statistics, there are all sorts of variations.

  A set of boxplots in the same window can be either boxplots for the same variable with one for each subgroup or boxplots for different variables. It is necessary to know which type you have in front of you. Boxplots by group must have the same scale and could be drawn with their width a function of the size of the group. Boxplots of different variables may have different scales and each case appears in each boxplot (apart from missing values), so that there is no need to consider different widths.

3.7   Modelling and testing for continuous variables

1. Means
   The most common test for continuous data is to test the mean in some way, either
   against a standard value, or in comparison to the means of other variables, or by
   subsets. Mostly a t-test is used. It would be invidious to select a reference here as
   there are so many texts covering the topic. Alternatively medians may be tested,
   especially in conjunction with using boxplots.

2. Symmetry
   [Zheng and Gastwirth, 2010] discusses several tests of symmetry about an unknown median and also proposes bootstrapping to improve the power of the tests.

3. Normality
   There are a number of tests for normality (e.g., Anderson-Darling, Shapiro-Wilk,
   Kolmogorov-Smirnov). These have low power for small samples and may be
   rather too powerful for really large samples. A large sample will tend to have
   some feature that will lead to rejection of the null hypothesis. There is a book
   on testing for normality [Thode Jr., 2002] and there is an R package **nortest**
   which offers five tests to add to the Shapiro-Wilk test offered in the **stats** package
   that comes with R. Tests assess the evidence as to whether there has been some
   departure from normality, while graphics, especially Q-Q plots, help identify the
   degree and type of departure from normality.

4. Density estimation
   There are too many packages in R which offer some form of density estimation
   or other for it to be possible to list them all. They fit density estimates, but do
   not test. Choose the one (or ones) that you think are good and use it (or them).
   Bear in mind that densities for variables with strict boundaries (e.g., no negative
   values) need special treatment at the boundaries. At least one of the R packages,
   logspline, offers an option for this problem. Most do not.

5. Outliers
   The classic book on outliers [Barnett and Lewis, 1994] describes many tests for
   outliers, mostly for univariate distributions and individual cases. How useful they
   may be depends on the particular application. As the book counsels, you need
   to watch out for both masking (one group of outliers prevents you recognising
   another) and swamping (mistaking standard observations for outliers).

6. Multimodality
   Good and Gaskin introduced the term 'Bump-Hunting' for looking for modes
   in an oft-cited article [Good and Gaskins, 1980]. The dip test for testing for unimodality was proposed in [Hartigan and Hartigan, 1985] and it is available in R
   in the appropriately named package **diptest**.

Main points

1. There are lots of different features that can arise in the frequency distributions of single continuous variables (e.g., Figure 3.9).

2. There is no optimal type of plot and no optimal version of a plot type. Look at several different plots and several different versions of each (cf. §3.3).

3. Natural binwidths based on context are usually a good choice for histograms (e.g., Figures 3.1 and 3.13).

4. Histograms are good for emphasising features of the raw data, while density estimates are better for suggesting underlying models for the data (e.g., Figures 3.5 and 3.10).

5. Boxplots are best for identifying outliers (Figure 3.12) and for comparing distributions across subgroups (Figure 3.14).

Exercises

1. **Galaxies**
   The dataset *galaxies* in the package **MASS** contains the velocities of 82 planets.

   (a) Draw a histogram, a boxplot, and a density estimate of the data. What information can you get from each plot?

   (b) Experiment with different binwidths for the histogram and different bandwidths for the density estimates. Which choices do you think are best for conveying the information in the data?

   (c) How many plots do you think you need to present the information? Which one(s)?

2. **Boston housing**
   The dataset is called *Boston* from the package **MASS**.

   (a) Figure 3.9 displays histograms of the 14 variables. How would you describe the distributions?

   (b) For which variables, if any, might boxplots be better? Why?

3. **Student survey**
   The data come from an old survey of 237 students taking their first statistics course. The dataset is called *survey* in the package **MASS**.

   (a) Draw a histogram of student heights and overlay a density estimate of the data. Is there evidence of bimodality?

   (b) Experiment with different binwidths for the histogram and different bandwidths for the density estimates. Which choices do you think are best for conveying the information in the data?

   (c) Compare male and female heights using separate density estimates that are common scaled and aligned with one another.

4. **Movie lengths**
   The *movies* dataset in the package **ggplot2** was introduced in §3.3.

   (a) Amongst other features we mentioned the peaks at 7 minutes and 90 minutes. Draw histograms to show whether these peaks existed both before and after 1980.

   (b) One variable, Short says whether a film was classified as a 'short' film ('1') or not ('0'). What plots might you draw to investigate which rule was used to define 'short' and whether the films have been consistently classified? (Hint: make sure you exclude the high outliers from your plots or you won't see anything!)

5. **Zuni educational funding**
   The *zuni* dataset in the package **lawstat** seems quite simple. There are three pieces of information about each of 89 school districts in the U.S. State of New Mexico: the name of the district, the average revenue per pupil in dollars, and

the number of pupils. This apparent simplicity hides an interesting story. The data were used to determine how to allocate substantial amounts of money and there were intense legal disagreements about how the law should be interpreted and how the data should be used. Gastwirth was heavily involved and has written informatively about the case from a statistical point of view, [Gastwirth, 2006] and [Gastwirth, 2008]. One statistical issue was the rule that before determining whether district revenues were sufficiently equal, the largest and smallest 5% of the data should first be deleted.

(a) Are the lowest and highest 5% of the revenue values extreme? Do you prefer a histogram or a boxplot for showing this?

(b) Having removed the lowest and highest 5% of the cases, draw a density estimate of the remaining data and discuss whether the resulting distribution looks symmetric.

(c) Draw a Q-Q plot for the data after removal of the 5% at each end and comment on whether you would regard the remaining distribution as normal.

6. **Non-detectable**
The dataset *CHAIN* from the package **mi** includes data from a study of 532 HIV patients in New York. The variable h39b.W1 records, according to the R help page, 'log of self reported viral load level at round 6th (0 represents undetectable level)'. Using the function table(CHAIN$h39b.W1) you can find out that there are 188 cases with value 0, i.e., with undetectable levels. Further examination of the dataset (using mi.info(CHAIN), for instance) reveals that additionally 179 cases have missing values for h39b.W1. What plots would you draw to show the distribution of the variable CHAIN$h39b.W1?

(a) with the 0 values?

(b) without the 0 values?

7. **Diamonds**
The set *diamonds* from the package **ggplot2** includes information on the weight in carats and price of 53,940 diamonds.

(a) Is there anything unusual about the distribution of diamond weights? Which plot do you think shows it best? How might you explain the pattern you find?

(b) What about the distribution of prices? With a bit of detective work you ought to be able to uncover at least one unexpected feature. How you discover it, whether with a histogram, a dotplot, a density estimate, or whatever, is unimportant, the important thing is to find it. Having found it, what plot would you draw to present your results to someone else? Can you think of an explanation for the feature?

8. **Intermission**
Albrecht Dürer's preparatory drawing *Praying Hands* is in the *Albertina Museum* in Vienna. Could the person whose hands are portrayed have worked several years in a mine? What does that say about the story of the origin of the picture circulating on the web?

4

Displaying Categorical Data

It's not what you look at that matters, it's what you see.

Thoreau

Summary

Chapter 4 looks at visualising single categorical variables, nominal, ordinal, or discrete.

4.1 Introduction

Compared to the range of plots proposed for visualising single continuous variable data, there have been few suggestions for visualising single categorical variable data. Barcharts (horizontally or vertically) and piecharts are the commonest options. You could argue that both are area plots where cases are represented by an area proportional to group size rather than being plotted individually. However, since the bars in a barchart always have the same width, you compare lengths, not areas, which is much easier.

Occasionally there are suggestions for using individual points for cases and jittering to keep them apart. This does not work well for high-frequency groups, as it is hard to assess their densities, and the displays for low-frequency groups may exhibit non-existent patterns due to the random jittering. Nevertheless, as always with exploratory graphics, if a graphic helps to uncover information, it is worth using.

Whether the graphic that is used to find information is the best way of presenting it to others is another matter. Single categorical variables usually contain little information, but it is still worth looking at some examples before moving on to consider multivariate categorical data and categorical variables together with continuous variables. There is often information in the simplest plots that is obscured in more complicated displays.

There are sixteen states (*Bundesländer*) in the German Federal Republic. Figure 4.1 shows three displays of the states using the number of eligible voters in the 2009 election as a proxy for size. The first display uses alphabetical order. Some states are much bigger than others, with *Nordrhein-Westfalen* being the biggest and *Bremen* (HB is *Hansestadt Bremen*) the smallest. The second display orders by size, and it is easier to judge the full ordering and see that *Schleswig-Holstein* (SH) is bigger than *Brandenburg* (BB). The final display orders by size within East/West groupings, where *Berlin* (BE) has been classed as 'East', as it is geographically. In fact, part of it, West Berlin, used to be part of West Germany. It is now clearer that the states in the East are all roughly the same size, with Sachsen being a bit bigger. Both the biggest and smallest states are in the West.

The data need some preparation: a new variable is created with abbreviated names for the *Bundesländer*; numbers of eligible voters are aggregated by *Bundesland* (the data are provided for the 299 constituencies); and a new variable is created according to whether the *Bundesländer* were part of the old West or East Germany.

```
data(btw2009, package = "flexclust")
btw2009 <- within (btw2009, stateA <- state)
btw2009 <- within (btw2009,
                   levels(stateA) <- c("BW", "BY", "BE",
                   "BB", "HB", "HH", "HE", "MV", "NI", "NW",
                   "RP", "SL", "SN", "ST", "SH", "TH"))
Voters <- with(btw2009, size <- tapply(eligible, stateA, sum))
Bundesland <- rownames(Voters)
btw9s <- data.frame(Bundesland, Voters)
btw9s$EW <- c("West")
btw9s[c("BB", "BE", "MV","SN","ST","TH"), "EW"] <- "East"
ls <- with(btw9s, Bundesland[order(EW, -Voters)])
btw9s <- within(btw9s, State1 <- factor(Bundesland, levels=ls))
```

The first version is best if you are used to that default order and know that, for instance, *Berlin* is always in second position. Except that it is not. With the full names of the states, rather than the standard abbreviations, you get a different ordering and *Berlin* is in third position. Consistent orderings are helpful for looking at displays of several different variables, you just have to agree on the ordering. The other versions are better for specific information relating to a single variable. Whichever one you use, they all give a quick overview of the relative sizes of the sixteen states. Barcharts are good for that. For those who would like an older example, take a look at the top plot on page 11 of [Gannett, 1898]. It shows a horizontal barchart of the "population of each state and territory" of the United States in 1890.

The German election data are an example of categorical data, counts by category. The categories may be nominal with no standard order (like a group of instant coffee brands) or ordinal (for instance when age is recorded as 'young', 'middle-aged', or 'old') or discrete (such as the number of children in a family). When the data are ordinal or discrete, the order must be preserved. When the data are nominal, reordering the categories becomes an important tool for gaining insights from the data.

```
b1 <- ggplot(btw9s, aes(Bundesland, Voters/1000000)) +
          geom_bar(stat="identity") +
          ylab("Voters (millions)")
b2 <- ggplot(btw9s, aes(reorder(Bundesland, -Voters),
          Voters/1000000)) + geom_bar(stat="identity") +
          xlab("Bundesland") + ylab("Voters (millions)")
b3 <- ggplot(btw9s, aes(State1, Voters/1000000)) +
          geom_bar(stat="identity")  + xlab("Bundesland") +
          ylab("Voters (millions)")
grid.arrange(b1, b2, b3)
```

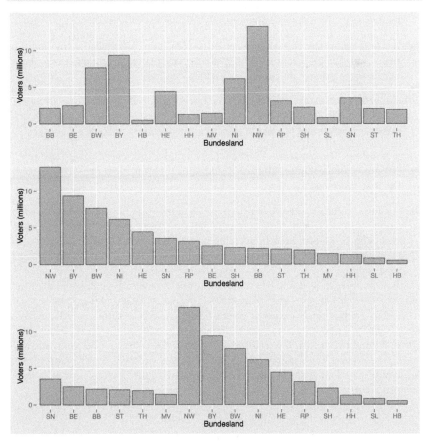

FIGURE 4.1: Three alternative barcharts of the numbers of eligible voters in the German *Bundesländer*. The top one is the default alphabetical order, the middle one is ordered by decreasing number of voters, and the bottom one is ordered by voter numbers within the former East/West Germanies. There is great variation in size, and the biggest *Bundesländer* are in the area that was formerly West Germany.

4.2 What features might categorical variables have?

There might be

Unexpected patterns of results There may be many more of some categories than others. Some categories may be missing completely.

Uneven distributions Observational studies may exhibit some form of bias, perhaps too many males. In medical meta analyses many trials are analysed together, although it can turn out that most of the trials were small and that the results are dominated by the one or two major trials.

Extra categories Gender may be recorded as 'M' and 'F', but also as 'm' and 'f', 'male' and 'female'. In a study of patients with two medical conditions, there may be some patients diagnosed with a third condition.

Unbalanced experiments Although experiments are usually carefully designed and carried out, there is always the chance that some data are missing or unusable. It is important to know if this occurs and leads to unequal group sizes.

Large numbers of categories In studies including open-ended questions (e.g., "Who is your favourite politician?") there may be far more names mentioned than you expected.

Don't knows, refusals, errors, missings, ... Data may not be available for a wide variety of reasons, and plots summarising how many cases of each type have arisen can be helpful both in deciding how to handle the data and in properly qualifying the results from the data that are available. Opinion polls are an obvious example.

4.3 Nominal data—no fixed category order

Meta analyses—how big was each study?

The package **meta** includes three meta analysis datasets. The dataset *Fleiss93* has details of seven studies on the use of aspirin after myocardial infarction. Figure 4.2 plots a barchart of the study sizes, with the studies ordered by the total numbers of patients (control group and experimental group together). The differences in size are striking, with one study being much bigger than all the others.

```
data(Fleiss93, package="meta")
Fleiss93 <- within(Fleiss93, {
                  total <- n.e + n.c
                  st <- reorder(study, -(total)) })
ggplot(Fleiss93, aes(st, total)) + geom_bar(stat="identity") +
      xlab("") + ylab("") + ylim(0,20000)
```

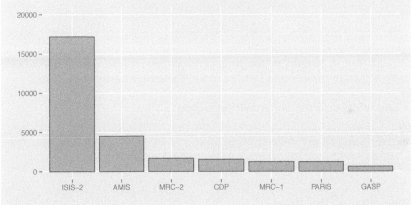

FIGURE 4.2: A barchart of the sizes of the seven studies in the *Fleiss93* meta analysis dataset. The ISIS-2 study had more patients than all the others put together.

If there were many more than seven studies, and that is the usual situation with meta analyses, it could make sense to plot individual bars for the biggest studies and a combined bar for the total of the smaller studies. The following code would draw the corresponding plot for *Fleiss93*, combining all studies with fewer than 2000 cases:

```
Fleiss93 <- within(Fleiss93, st1 <- as.character(study))
Fleiss93$st1[Fleiss93$total < 2000] <- "REST"
ggplot(Fleiss93, aes(st1, total)) + geom_bar(stat="identity") +
      xlab("") + ylab("") + ylim(0,20000)
```

Anorexia

In the *anorexia* dataset in the **MASS** package, two treatment groups are compared with a control group. There are 72 cases and you might assume that they were split equally into three groups of 24 each. Figure 4.3 shows the curiously uneven distribution (29 and 17 for the two treatment groups and 26 for the control group). Needless to say, there were probably very good reasons for the split, but it could be useful to know. What if the groups were initially of the same size and there were dropouts?

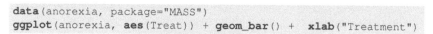

```
data(anorexia, package="MASS")
ggplot(anorexia, aes(Treat)) + geom_bar() + xlab("Treatment")
```

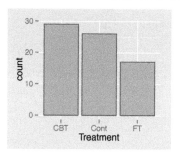

FIGURE 4.3: A barchart of the group sizes in the *anorexia* dataset drawn with `barplot(table(anorexia$Treat))`. The groups are not of equal size.

A table would show this just as well in principle

```
with(anorexia, table(Treat))
```

but a graphic is more convincing.

Who sailed on the Titanic?

The tragic sinking of the Titanic has been discussed endlessly and there are many films, books, and webpages about the disaster. Despite all this attention the information on those who sailed on the ship is incomplete. A lot is known about some passengers and members of the crew, especially about the first-class passengers and the officers; less is known about some of the others on the ship.

Although there is not full agreement on all details, the dataset *Titanic* is generally thought to be an accurate summary. It contains four pieces of information on 2201 people who were on board: `Class` (1st, 2nd, 3rd, or crew), `Sex`, `Age` (a binary variable, Child or Adult), and `Survived` (whether they survived or died). Interest centres on issues such as survival by `Class` and `Sex` (which will be discussed in §7.2), but it is sensible to have a look at the simple univariate barcharts first, as in Figure 4.4.

```
Titanic1 <- data.frame(Titanic)
p <- ggplot(Titanic1, aes(weight=Freq)) +
        ylab("") + ylim(0,2250)
cs <- p + aes(Class) + geom_bar(fill="blue")
sx <- p + aes(Sex) + geom_bar(fill="green")
ag <- p + aes(Age) + geom_bar(fill="tan2")
su <- p + aes(Survived) + geom_bar(fill="red")
grid.arrange(cs, sx, ag, su, nrow=1, widths=c(3, 2, 2, 2))
```

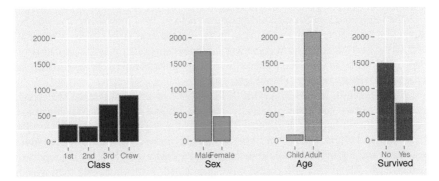

FIGURE 4.4: Barcharts of the four variables in the *Titanic* dataset. That the majority of passengers and crew died and that there were far more men than women on board is well known, the numbers in the different classes less so.

You should always think about what you expect graphics to show, before you draw them. That way you can be surprised by what you see and value more the information presented. Once you have seen Figure 4.4, it seems obvious that there were more people in the crew than in any of the other classes and that there were fewest in the second class. That there were more males than females will have surprised no one, but that there were over three times as many does seem high, probably because the crew were almost all men. The small number of children may raise questions about the quality of the data for the `Age` variable. The fact that overall twice as many died as survived is well known.

There are technical points worth mentioning. The dataset is supplied as four conditional contingency tables, so it has been converted to a dataframe. A common vertical scale has been used for all four plots to make comparisons across plots possible. This is fine for the three binary variables, but not so good for the variable `Class`, a typical issue when categorical variables have different numbers of categories. (The upper limit was chosen after drawing default plots. More formally, the maximum should be determined by calculation.)

The `Class` barchart has been drawn wider than the others, as it has more categories. This makes the code more complicated, but it is good to have the flexibility to do it. For presentation purposes it would have been more elegant to make all bars the same width. For exploratory purposes this would be gilding the lily.

Opinion polls

Nowadays we are bombarded with results of opinion polls on all kinds of subjects, and political opinion polls get special emphasis for obvious reasons. Although there are always people who fail to understand the question, who do not want to choose any of the options offered, or who just refuse to answer pollsters' questions, the numbers of 'Don't knows', or 'Undecideds' or whatever you choose to call them, are seldom reported. For political polls this group can be important, as they could sway the election if they did indeed decide to vote.

The referendum on Scottish independence held in 2014 looked like it would be very close just before voting took place. In the end the margin was fairly substantial (55.3% to 44.7%). Afterwards some commentators suggested that this should have been predictable, as the 'Don't knows' in opinion polls in referenda have tended to vote for the status quo [FiveThirtyEight, 2014].

A poll taken in Ireland in August 2013 shows how much difference neglecting the 'Don't knows' can make. Unusually, perhaps because of the relatively high proportion of 'Don't knows', the Sunday Independent newspaper presented the results both with and without the 'Don't knows' [Sunday Independent, 2013]. They used piecharts, and Figure 4.5 shows something similar drawn with R using the reported percentages. Altogether, 985 people were questioned, 4 of whom said they would vote for the Green party. Rounding the percentages, as was done by the newspaper, meant that the Green party was omitted completely from the published piechart with the 'Don't knows', but was shown in the second piechart.

The top plot, including the 'Don't knows', emphasizes how open the election was likely to be, whereas the lower plot suggests that Fianna Fail and Sinn Fein might have been able to form a government together (should such a coalition have been in the realms of possibility). Piecharts are good for emphasizing when data are shares of some whole, and can be helpful, as here, for helping to estimate possible coalition shares. You have to know what the shares are shares of. In the upper plot it is shares of all respondents, in the lower one it is shares of respondents who said whom they would support. Piecharts are poor for comparing values. Can you tell whether Fine Gael or Fianna Fail had more support in the survey? As the two major parties historically, it is very important to them which one is ahead. In this case, Fine Gael edged it by around 1%.

The piecharts could probably be improved with a better choice of colour shades and something close to the standard party colours should be chosen. It could be helpful to also report the values, either on the plots or in accompanying tables. Piecharts are not generally recommended (see, for example, the R help page for piecharts, where Bill Cleveland is quoted on the comparatively poor accuracy of judgements of angles) and should mostly be avoided. Piecharts in 3D should most definitely always be avoided.

```
Party <- c("Fine Gael", "Labour", "Fianna Fail",
           "Sinn Fein", "Indeps", "Green", "Don't know")
nos <- c(181, 51, 171, 119, 91, 4, 368)
IrOP <- data.frame(Party, nos)
IrOP <- within(IrOP, {
             percwith <- nos/sum(nos)
             percnot <- nos/sum(nos[-7])})
par(mfrow=c(2,1), mar = c(2.1, 2.1, 2.1, 2.1))
with(IrOP, pie(percwith, labels=Party, clockwise=TRUE,
           col=c("blue", "red", "darkgreen", "black",
           "grey", "lightgreen", "white"), radius=1))
with(IrOP, pie(percnot[-7], labels=Party, clockwise=TRUE,
           col=c("blue", "red", "darkgreen", "black",
           "grey", "lightgreen"), radius=1))
```

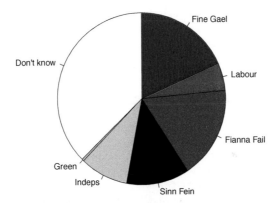

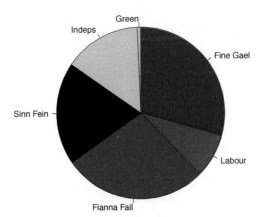

FIGURE 4.5: Piecharts of an Irish political opinion poll from August 2013, one including the 'Don't knows' and one not. The top plot emphasizes the large number of voters who have not decided how they might vote, while the second plot suggests that a Fianna Fail and Sinn Fein coalition, were such an agreement possible, could almost have a majority.

4.4 Ordinal data—fixed category order

Surveys

In many surveys there is a string of questions to which respondents give answers on integer scales from, say, 1 to 5. Sometimes descriptive terms are attached to each number on the scale, although not always.

The *BEPS* dataset in the **effects** package includes seven such questions put to 1525 voters in the British Election Panel Study for 1997-2001. One question was on a scale from 1 to 11, one from 0 to 3, and the rest were from 1 to 5. Figure 4.6 shows the barcharts for the voters' assessment of the leaders of the three main parties at the time, where a higher value means a better assessment. Few voters used the extremes of 1 or 5. The leaders of the two main parties, Tony Blair and William Hague, were either liked (a value of 4 or 5) or disliked (a value of 1 or 2). The middle value of 3 was hardly used for them, although it was used for Charles Kennedy, the leader of the Liberals. Once you have seen the patterns, there is no difficulty in describing them and suggesting possible explanations, but would you have expected them?

```
data("BEPS", package="effects")
a1 <- ggplot(BEPS, aes(factor(Hague))) +
            geom_bar(fill="blue") + ylab("") +
            xlab("Hague (Conservative)") + ylim(0, 900)
a2 <- ggplot(BEPS, aes(factor(Blair))) +
            geom_bar(fill="red") + ylab("") +
            xlab("Blair (Labour)") + ylim(0, 900)
a3 <- ggplot(BEPS, aes(factor(Kennedy))) +
            geom_bar(fill="yellow") + ylab("") +
            xlab("Kennedy (Liberal)") + ylim(0, 900)
grid.arrange(a1, a2, a3, nrow=1)
```

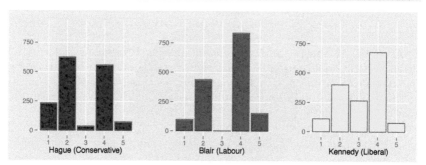

FIGURE 4.6: Assessment of UK party leaders by voters in 2001. Higher values mean a higher rating. The bars have been filled with the party colours. Opinion seems to be divided on the leaders of the two main parties, Hague and Blair.

Two of the variables concerned Europe, one the respondents' knowledge of the parties' policies on European integration and one the respondents' attitudes towards European integration. What patterns of results might be expected here? Figure 4.7 displays the two barcharts.

Well over half the respondents thought they had good or very good knowledge of party policies (options 2 and 3), while a sizeable minority thought they had low knowledge (option 0). Very few chose the option 1. The second barchart plots a scale based on combining answers to several questions, so it is reasonable to find that there is a spread of scale values, although that they are represented fairly equally seems a little surprising. The biggest group was the over 20% who were strongly against European integration. Given the tenor of the discussions about membership of the European Union going on in the UK in 2014, the proportion strongly against European integration now is probably considerably higher than it was at the turn of the century.

```
b1 <- ggplot(BEPS, aes(factor(political.knowledge))) +
        geom_bar(fill="tan2")  + coord_flip() + ylab("") +
        xlab("Knowledge of policies on Europe")
b2 <- ggplot(BEPS, aes(factor(Europe))) +
        geom_bar(fill="lightgreen") + ylab("") +
        xlab("Attitudes to European integration")
grid.arrange(b1, b2, nrow=1, widths=c(4, 8))
```

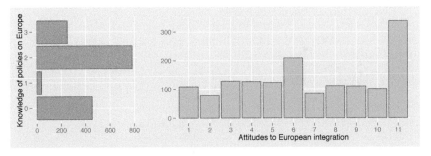

FIGURE 4.7: UK voters' attitudes to European integration are shown in the plot on the right. There is a strongly negative group (those with the value '11'), but otherwise the respondents are roughly equally spread across the scale. The barchart on the left referring to respondents' knowledge has been drawn horizontally to differentiate it from the barchart on the right referring to respondents' opinions.

And more surveys

The *survey* dataset from the **MASS** package gives the results of a survey of 237 Australian statistics students. (It is also used in Exercise 3 of Chapter 3.) Figure 4.8 displays the barcharts for the seven categorical variables and gives an idea of the range of barcharts that you can find.

For most of the variables a simple line of code suffices, adding informative labels instead of the variable names. The variables for exercising and smoking have had to be reordered for the plots and this is something that always has to be watched out for. Note that the vertical scales vary from about 120 to just over 200.

Interestingly the sexes look exactly equal (a table shows they were), and one person did not give that information. Most students are right-handed (again one person did not answer the question, not the same one as before, as it turned out). More students fold their arms right over left, and a non-negligible minority say that neither right nor left is on top. With clapping, the majority for the right side is higher and the 'neither' group is bigger. As you might expect, most students do at least some exercising and few smoke. Many people in Australia still use Imperial units (feet and inches) to measure height, or rather they did at the time the survey was carried out in the 1970s.

```
data(survey, package="MASS")
s1 <- ggplot(survey, aes(Sex)) + geom_bar() + ylab("")
s2 <- ggplot(survey, aes(W.Hnd)) + geom_bar() +
            xlab("Writing hand") + ylab("")
s3 <- ggplot(survey, aes(Fold)) + geom_bar() +
            xlab("Folding arms: arm on top") + ylab("")
s4 <- ggplot(survey, aes(Clap))  + geom_bar() +
            xlab("Clapping: hand on top") + ylab("")
survey <- within(survey,
                 ExerN <- factor(Exer,
                 levels=c("None", "Some", "Freq")))
s5 <- ggplot(survey, aes(ExerN))  + geom_bar() +
            xlab("Exercise") +  ylab("")
s6 <- ggplot(survey, aes(M.I))  + geom_bar() +
            xlab("Height units") + ylab("")
survey <- within(survey,
                 SmokeN <- factor(Smoke,
                 levels=c("Never", "Occas", "Regul", "Heavy")))
s7 <- ggplot(survey, aes(SmokeN))  + geom_bar() +
            xlab("Smoking") + ylab("")
grid.arrange(s1, s2, s3, s4, s5, s6, s7, ncol=3)
```

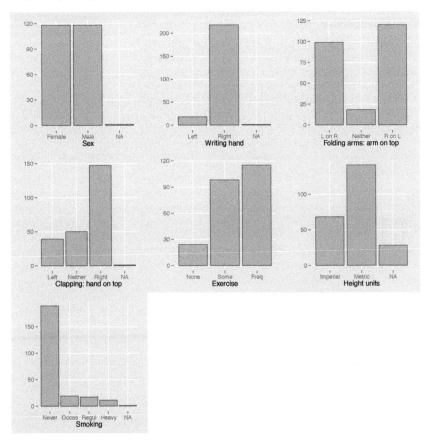

FIGURE 4.8: Barcharts of the categorical variables in the *survey* dataset of 237 statistics students. The individual plots are discussed in the chapter. Barcharts may be of many different shapes. Each plot has its own vertical axis scale.

4.5 Discrete data—counts and integers

Deaths by horsekicks

Von Bortkiewicz's dataset about deaths by horsekick in the Prussian army over
twenty years is included in **vcd** under the name *vonBort*. Figure 4.9 shows the distribution of the numbers of deaths by corps and year. There is no information provided
on the sizes of the corps over the years, though it is believed that most had four regi-
ments, while three had six and one had eight. There may have been around 750 men
(and horses) in each cavalry regiment.

```
data(VonBort, package="vcd")
ggplot(VonBort, aes(x=factor(deaths))) + geom_bar() +
       xlab("Deaths by horse kick")
```

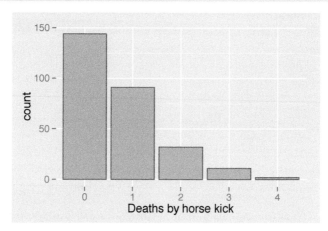

FIGURE 4.9: Number of soldiers killed in one of 14 Prussian army corps by year for
twenty years from 1875 to 1894. The distribution is close to a Poisson distribution
with parameter 0.7, suggesting that these might be considered unfortunate random
accidents. The dataset first appeared in [von Bortkiewicz, 1898].

A test of whether a Poisson distribution would be a good fit is not significant.

```
gf <- goodfit(table(VonBort$deaths))
summary(gf)

#
#       Goodness-of-fit test for poisson distribution
#
#                        X^2 df  P(> X^2)
#    Likelihood Ratio 2.442786  3 0.4857196
```

Looking at the distributions for individual years illustrates a problem that can arise with data like these. Both plots in Figure 4.10 show the distribution of deaths by corps for the year 1891. No corps suffered either 2 or 4 deaths and so the plot on the left ignores those categories. This can be misleading for categories outside the range of the subset (the value 4 in this case) and is more serious for a missing category inside the range (the value 2 here). The plot on the right solves the problem by specifying the levels of the factor deaths.

```
data(VonBort, package="vcd")
h1 <- ggplot(VonBort[VonBort$year=="1891", ],
             aes(x=factor(deaths))) + geom_bar()
h2 <- ggplot(VonBort[VonBort$year=="1891", ],
             aes(x=factor(deaths, levels=seq(0, 4)))) +
             geom_bar() + scale_x_discrete(drop=FALSE)
grid.arrange(h1, h2, nrow=1)
```

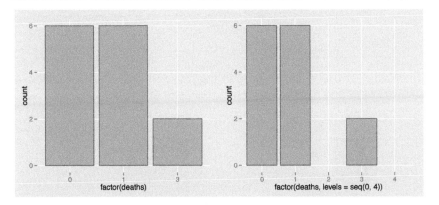

FIGURE 4.10: Numbers of soldiers killed in one of 14 Prussian army corps in 1891. The plot on the left leaves out values with zero counts, the plot on the right is a more complete display. With only 14 pieces of data, it is not surprising that there can be two unusual values (unusual for the year 1891).

Goals in soccer

Scores in many sports lead to discrete data, whether goals in soccer, runs in cricket, points in rugby, or shots in golf. The dataset *UKSoccer* in **vcd** summarizes the results of season 1995-96 of the English Premier League in a table giving the frequencies with which specific results occurred. The second line of code converts the table into a data frame and Figure 4.11 shows the distributions of goals scored by the home and away teams. If a team scored more than 4 goals, 4 was recorded in the dataset, which is why the labels have been changed in the code. The vertical scales have been made the same for comparative purposes.

```
data(UKSoccer, package="vcd")
PL <- data.frame(UKSoccer)
lx <- c("0","1","2","3","4 or more")
b1 <- ggplot(PL, aes(x=factor(Home), weight=Freq)) +
            geom_bar(fill="firebrick1") +
            ylab("") + xlab("Home Goals")  +
            scale_x_discrete(labels=lx)  + ylim(0,150)
b2 <- ggplot(PL, aes(x=factor(Away), weight=Freq)) +
            geom_bar(fill="cyan1") +
            ylab("") + xlab("Away Goals")  +
            scale_x_discrete(labels=lx)  + ylim(0,150)
grid.arrange(b1, b2, nrow=1)
```

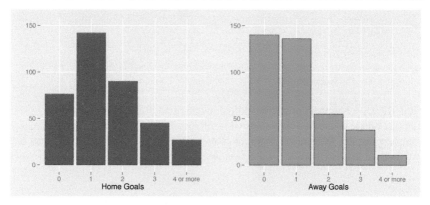

FIGURE 4.11: Distributions of the numbers of goals scored by the home and away teams in the 1995-96 English Premier League season. As you would expect, the home teams scored more goals. Perhaps it is unexpected that there are relatively few matches in which either team scores more than two goals.

Benford's Law

The observation that the first or leading digits of many sets of numbers do not follow a uniform distribution was known a long time ago. As Simon Newcomb wrote in his 1881 article [Newcomb, 1881]: "That the ten digits do not occur with equal frequency must be evident to any one making much use of logarithmic tables, and noticing how much faster the first pages wear out than the last ones." However, the result is known as Benford's Law, because of his article [Benford, 1938], which described the checking of many real datasets he had collected.

There have been several interesting applications in recent years, including studies of financial fraud and voting irregularities, and references can be found in the relevant Wikipedia article. The package **benford.analysis** provides six datasets, including *census.2009*, which gives the populations of 19,509 U.S. towns and cities.

If we only consider the first digit, then there are only nine possible values and the probability distribution is shown in Figure 4.12:

```
xx <- 1:9
Ben <- data.frame(xx, pdf=log10(1+1/xx))
ggplot(Ben, aes(factor(xx), weight=pdf)) + geom_bar() +
      xlab("") + ylab("") + ylim(0,0.35)
```

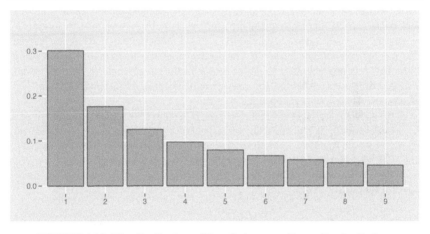

FIGURE 4.12: The distribution of first digits according to Benford's law.

4.6   Formats, factors, estimates, and barcharts

Attentive readers will have noticed that drawing graphics of categorical data often involves preparatory reformatting of the data. Working with categorical data is not always straightforward and you have to watch both for how the data are stored and how the categories are coded.

Shape of the dataset

The data may be provided as a collection of individual cases, as a list of the possible combinations of the categorical variables with their associated frequencies, or as a set of conditional tables. Depending on which form the data have, some restructuring may be necessary before drawing particular plots.

Coding of variables

Categorical variable data may be recorded in different ways. Numeric codings were common in the distant past when computer storage was expensive and limited. Numeric codings have their uses, but there is always the danger of inappropriate calculations being carried out (e.g., what is the average marital status when '1' = single, '2' = married, '3' = separated, ... ?), and the implied ordering may need to be changed too. You can convert the variable to a factor to change the order, but if you are going to do that, it is probably best to define a new factor variable and replace the codes with descriptive text. This has the advantage of keeping the original variable: you never know when you might need it.

If the data are stored as character strings, R assumes the variable is a factor and the default ordering is alphabetical. Non-alphabetical orderings can be specified using the `levels` function. Sometimes this has been done already for datasets available in R, and you may have to change the ordering, if the default is not what you need. Reordering should always be carried out with care, since there are too many ways for things to go wrong.

Estimates shown as bars

In some fields of application estimates are displayed as bars. When confidence intervals are shown as well, the lower part can be hidden by the colouring of the bar and the upper part looks like the plunger on a dynamite box, which gives rise to the name 'dynamite plot'. Most statisticians disapprove and recommend using dotplots with confidence intervals instead [Cleveland, 1993]. In this book there are no dynamite plots, only barcharts where the bars represent counts or probabilities.

4.7 Modelling and testing for categorical variables

1. Testing by simulation
χ^2 tests in one form or another are standard. An alternative to relying on the asymptotic distribution of the test statistic is to simulate many datasets and compare the actual test statistic value with the distribution of simulated values. This option is available in the R function `chisq.test`.

2. Evenness of distribution
If numbers are supposed to have been drawn at random, for instance in a lottery, or if a random number generator is to be checked directly, then the null hypothesis of equally likely probabilities can be carried out with a χ^2 test.

3. Fitting discrete distributions
As always with categorical data, χ^2 tests are important, but it is useful to inspect any lack of fit visually. Tukey's hanging rootograms are one approach; plotting components of χ^2-statistics is another.

Main points

1. Barcharts are a simple form of display, yet they can provide much information and often surprise. Errors may become apparent, there may be more categories than anticipated, category counts may be unexpected (e.g., Figures 4.2 and 4.6).

2. Barcharts can be used for nominal, ordinal, or discrete variables.

3. The order of categories affects the way a barchart of a nominal variable looks, and different orderings can emphasise different features, as was shown in Figure 4.1.

4. People have strong opinions about piecharts. Don't let that stop you using them if you like them, just make sure they show clearly what you want them to. They are useful for displaying shares, as in Figure 4.5.

Exercises

1. **Gastrointestinal damage**
   The dataset is called *Lanza* from the package **HSAUR2**.

   (a) Data on four studies are reported. Draw a plot to show whether all four studies are equally large.
   (b) The outcome is measured by the variable `classification` with scores of 1 (best) to 5 (worst). How would you describe the distribution?

2. **Alzheimer's**
   Three groups of patients (one with Alzheimer's, one with other forms of dementia, and a control group with other diagnoses) were studied. Counts are given in the dataset *alzheimer* in the package **coin**. Prepare a graphical summary of plots of each of the three variables, `smoking`, `disease`, and `gender`, in a single window.

   (a) Are the disease groups very different in size?
   (b) Are there more men or women in the study?
   (c) How would you describe the distribution of the smoking variable? Do you think the smoking data are likely to be reliable?

3. **Slot machines**
   According to the R help entry, this dataset (*vlt* in package **DAAG**) was collected from the three windows of a video lottery terminal while playing the game 'Double Diamond'. There are seven possible symbols that may appear in each window. Draw equally scaled barcharts to see if the distributions of frequencies are the same for each window. Describe any important features.

4. **Multiple sclerosis**
   The dataset *MSPatients* in **vcd** provides information on the diagnoses of two neurologists from two different cities on two groups of patients, one from each city. How do the distributions of the ratings of the neurologists compare? How would you describe their rating patterns? Draw two barcharts with common scaling. Before drawing the barcharts, reorder the categories into a sensible ordinal scale instead of the default alphabetical order. (Hint: initially using `data.frame(as.table(MSPatients))` will put the dataset into a form that is easier to work with.)

5. **Occupational mobility**
   According to the R help page, the *Yamaguchi87* dataset in **vcdExtra** has become a classic for models comparing two-way mobility tables.

   (a) How do the distributions of occupations of the sons in the three countries compare?
   (b) How do the distributions of the sons' and fathers' occupations in the UK compare?
   (c) Are you surprised by the results or are they what you would have expected?

6. **Whisky**
   The package **bayesm** includes the dataset *Scotch*, which reports which brands of whisky 2218 respondents consumed in the previous year.

   (a) Draw a barchart of the number of respondents per brand. What ordering of the brands do you think is best?
   (b) There are 20 named brands and a further category Other.brands. That entails drawing a lot of bars. If you decided to plot only the biggest brands individually and group the rest all together in the 'Other' group, what cut-off would you use for defining a big brand?
   (c) Another version of the dataset called *whiskey* is given in the package **flexmix**. It is made up of two data frames, *whiskey* with the basic data, and *whiskey_brands* with information on whether the whiskeys are blends or single malts. How would you incorporate this information in your graphics, by using colour, by using a different ordering, or by drawing two graphics rather than one?
   (d) Which of the spellings, 'whisky' or 'whiskey', is more appropriate for this dataset?

7. **Choice of school**
   The dataset *GSOEP9402* in the package **AER** provides data on 675 14-year-old children in Germany. The data come from the German Socio-Economic Panel for the years 1994 to 2002.

   (a) Which variables are nominal, ordinal, or discrete?
   (b) Draw barcharts for the variables. Are any similar in form, and what explanations would you suggest for these similarities?
   (c) The variable meducation refers to the mother's educational level in years. Would you describe it as ordinal or discrete, and how should it be displayed?
   (d) Summarise briefly the main information shown by your graphics.

8. **Election results**
   The Bavarian election of Autumn 2013 was a triumph for the CSU party, as they obtained an absolute majority for the first time for several years. This was partly because some parties failed to get 5% of the votes or more and could therefore not win any seats in parliament. A few observers commented that the real winners were the group who did not vote at all, as they made up a higher proportion of the electorate than the CSU supporters. The percentages reported were as follows (where the party percentages refer to their share of the actual vote): 'didn't vote' 36.7%, CSU 47.7%, SPD 20.6%, FW 9.0%, Grüne 8.6%, FDP 3.3%, BP 2.1%, Linke 2.1%, ÖDP 2.0%, Pirates 2.0%, Rest 2.5%.

   (a) Draw graphics to present these results both with and without the group of non-voters. What headlines would you give your graphics if you were to publish them in a newspaper?
   (b) It is, of course, not necessary to list all the parties, especially when several of them were excluded by the 5% rule. Which ones would you leave out and why?
   (c) Seats are won by the parties with over 5% of the vote, based on their share of the total votes of qualifying parties. The rules are actually more complicated than that, but these complications can be neglected here. There are 180 seats in all. Draw a graphic showing the seat distribution, using the available figures. What headline would you suggest for this graphic?
   (d) You have probably noticed that the percentages do not add up to 100%. Is this a problem and what might you do about it?

9. **Horsekicks**
   In his discussion of the deaths due to horsekicks in the Prussian army, von Bortkiewicz pointed out that four of the corps (G, I, VI, and XI) were bigger than the others.

   (a) Draw two plots, one for the numbers of deaths each year for the four bigger corps and one for the other corps. What vertical scale do you think is appropriate for comparing the two plots?
   (b) Estimate the parameter of a Poisson distribution for the other corps. Carry out a test to see if a Poisson distribution with that λ would be an acceptable fit for the four bigger corps.

10. **Intermission**
    Edward Hopper's *Nighthawks* hangs in the *Art Insitute of Chicago*. What are the main features of this painting and is it typical of Hopper's work?

5

Looking for Structure: Dependency Relationships and Associations

'The world is full of obvious things which nobody by any chance ever observes.'

Sherlock Holmes (in Sir A. Conan Doyle's *The Hound of the Baskervilles*)

Summary

Chapter 5 is about examining how pairs of continuous variables are related.

5.1 Introduction

Drawing scatterplots is one of the first things statisticians do when looking at datasets. Scatterplots can reveal structure that is not readily apparent from summary statistics or from models, and they are relatively easy to understand and present. They are the basis of the Gapminder displays [Rosling, 2013] used so effectively by Hans Rosling to draw attention to patterns of world development.

The major role of scatterplots lies in revealing associations between variables, not just linear associations, but any kind of association. Scatterplots are also useful for identifying outliers and for spotting distributional features. However, marginal distributions cannot always be clearly seen from scatterplots, so this book recommends you always take a look at the univariate distributions briefly as well.

Over ten thousand athletes competed in the London Summer Olympics of 2012. Figure 5.1 shows a scatterplot of `Weight` against `Height` from the dataset *oly12* in the package **VGAMdata**. Note that 1346 of the 10,384 athletes in the dataset are not displayed in the plot because one or other of the two measurements is missing. The expected relationship between weight and height can be clearly seen, although it is a little obscured by some outliers, which distort the scales, and drawing a second plot with tighter limits would show the bulk of the data better. There is evidence of discretisation in the height measurements (notice the parallel vertical lines) and the same effect would be visible for the weight measurements but for the outliers. Given the large number of points, there is also a lot of overplotting, and most of the points in the middle of the plot represent more than one case (there are 57 athletes who are 1.7m in height and weigh 60kg). You can alleviate the problem using alpha-blending, giving each point a weight equal to the parameter `alpha` in `geom_point`, although outliers are then less easy to see.

```
data(oly12, package="VGAMdata")
ggplot(oly12, aes(Height, Weight)) + geom_point() +
      ggtitle("Athletes at the London Olympics 2012")
```

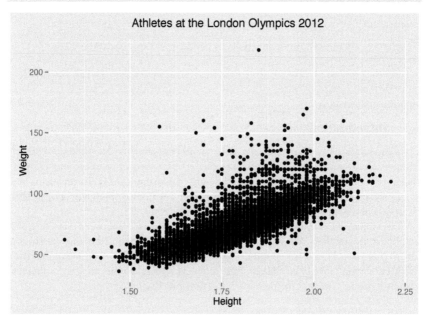

FIGURE 5.1: A scatterplot of weight against height for the athletes at the London Olympics 2012. You can see that weight increases with height. There are several outliers which affect the scales.

5.2   What features might be visible in scatterplots?

Features that might be found in a scatterplot include

Causal relationships (linear and nonlinear)   One variable may have a direct influence on another in some way. For example, people with more experience tend to get paid more. It is standard to put the dependent variable on the vertical axis. Sometimes initial ideas are misleading and X is more sensibly seen as depending on Y and not the other way round.

Associations   Variables may be associated with one another without being directly causally related. Children get good marks in English and in Maths because they are intelligent, not because ability in one subject is the reason for the ability in the other.

Outliers or groups of outliers   Cases can be outliers in two dimensions without being outliers in either dimension separately. Taller people are generally heavier, but there may be people of moderate height who are so heavy or light for their height that they stand out in comparison with the rest of the population.

Clusters   Sometimes there are groups of cases which are separate from the rest of the data. In a scatterplot of the two petal measurements for Fisher's iris dataset, Figure 1.4, you can see that the setosa flowers have much lower values than the other two varieties.

Gaps   Occasionally, particular combinations of values do not occur. Movies which are rated highly are rated often, but movies which few people like are seldom rated, so in a plot of average rating against number of ratings the bottom right of the plot is empty, Figure 5.7.

Barriers   Some combinations of values may not be possible. No one can have more years of experience than their age, and a plot of the two variables will have a linear boundary, which should not be crossed.

Conditional relationship   Sometimes the relationship between two variables is better summarised by a conditional description than by a function. A plot of income against age is likely to differ before and after retirement age.

The following two sections show a range of different scatterplots for a number of different applications.

5.3   Looking at pairs of continuous variables

The evils of drink?

Karl Pearson investigated the influence of drink on various aspects of family life at
the beginning of the twentieth century. The dataset *DrinksWages* is in the collection
of historic datasets made available in the package **HistData**. Figure 5.2 shows a
scatterplot of `wage` (average weekly wage in shillings) plotted against `drinks/n`
(the proportion of drinkers) for 70 different trades. There is no apparent relationship.

```
data(DrinksWages, package="HistData")
ggplot(DrinksWages, aes(drinks/n, wage)) + geom_point() +
       xlab("Proportion of drinkers") + xlim(0,1) + ylim(0,40)
```

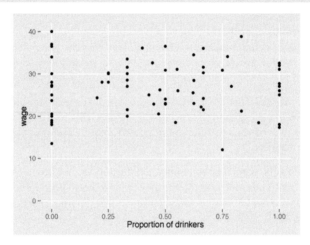

FIGURE 5.2: A scatterplot of average wages against proportion of drinkers for all
70 trades in Pearson's *DrinksWages* dataset in the **HistData** package. There are a
surprising number of trades where either all workers are drinkers or none are.

It is perhaps initially surprising that some trades have 100% drinkers and some
100% non-drinkers. A look at the distribution of numbers in the study by trade (or
indeed a table) explains why, as you can check for yourself using one of these alternatives:

```
with(DrinksWages, hist(n, breaks=0:max(n)))
with(DrinksWages, table(n))
```

Of the 70 trades, over a third (26) have only 1 or 2 members in the survey. The
biggest group with 100% drinkers (7 drinkers) turns out to be the chimney sweeps
and the biggest temperance group the gasworkers (5 non-drinkers):

```
with(DrinksWages, max(n[drinks==0]))
with(DrinksWages, trade[drinks==0 & n==max(n[drinks==0])])
with(DrinksWages, max(n[sober==0]))
with(DrinksWages, trade[sober==0 & n==max(n[sober==0])])
```

Excluding the smaller trades (say, all with less than five) gives Figure 5.3. There is no particular pattern and certainly no evidence of drink being associated with lower average wages. (Whether this is the best way to investigate this question is another matter.) Note that the scales in Figures 5.2 and 5.3 have been explicitly made equal to keep them comparable.

```
bigDW <- filter(DrinksWages, n > 4)
ggplot(bigDW, aes(drinks/n, wage)) + geom_point() +
        xlab("Proportion of drinkers") + xlim(0,1) +    ylim(0,40)
```

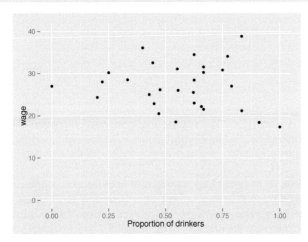

FIGURE 5.3: A scatterplot of average wages against proportion of drinkers for all trades with a group size of more than 4. There is no obvious relationship between the two.

Old Faithful

The Old Faithful geyser in Yellowstone National Park in Wyoming is a famous tourist attraction. The dataset *geyser* in **MASS** provides 299 observations of the duration of eruptions and the time to the next eruption. Figure 5.4 plots the data.

```
data(geyser, package="MASS")
ggplot(geyser, aes(duration, waiting)) + geom_point()
```

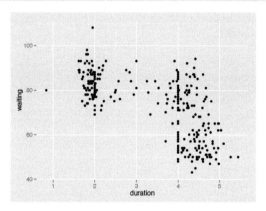

FIGURE 5.4: A scatterplot of the waiting time to the next eruption vs. the duration of the current eruption for the Old Faithful geyser in Yellowstone National Park. After an eruption of short duration, you have to wait longer for the next one.

A short duration implies a long waiting time until the next eruption, while a long duration can imply a short or long waiting time. There may be 3 clusters and possibly a couple of outlying values, but there is also a suggestion of rounded values for the eruption durations (note the numbers of durations of 2 and 4 minutes). To assess the possibility of clustering, consider a density estimate. Figure 5.5 displays contours of a bivariate density estimate supporting the idea of there being three clusters.

Contour plots of density estimates show equal levels of the estimated density function, but are not associated with probabilities. Using the **hdrcde** package you can estimate highest density regions, the smallest areas not including specified proportions of the distribution. Figure 5.6 displays the geyser data in this way for the proportions 1%, 5%, 50%, 75%. This suggests pretty much the same outliers as before, but does not support the three concentrations conclusion as strongly.

The original source of this dataset is not given on the R Help page. The age of the dataset and the apparently rounded values suggest that better quality data might be obtainable now. The Wikipedia page on Old Faithful is not consistent with these data at all [Wikipedia, 2013]:

> *With a margin of error of 10 minutes, Old Faithful will erupt 65 minutes after an eruption lasting less than 2.5 minutes or 91 minutes after an eruption lasting more than 2.5 minutes.*

```
ggplot(geyser, aes(duration, waiting)) + geom_point() +
    geom_density2d()
```

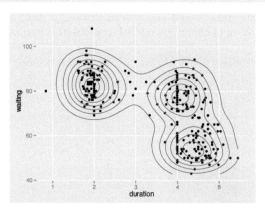

FIGURE 5.5: The same scatterplot with bivariate density estimate contours. There is evidence of three concentrations of data, two univariate outliers (one eruption with low duration and one with a high waiting time until the next eruption), and one bivariate outlier.

```
library(hdrcde)
par(mar=c(3.1, 4.1, 1.1, 2.1))
with(geyser, hdr.boxplot.2d(duration, waiting,
    show.points=TRUE, prob=c(0.01,0.05,0.5,0.75)))
```

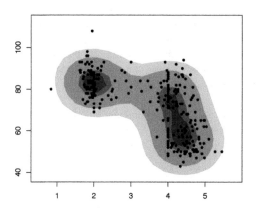

FIGURE 5.6: Another version of the scatterplot, but now with highest density regions based on a bivariate density estimate. There is less evidence of three data concentrations than in the previous plot and there is a slightly different set of possible outliers.

Movie ratings

The distribution of movie lengths in the dataset *movies* from **ggplot2** was discussed in §3.3. Here we are going to look at two other variables, `rating` (average IMDb user rating) and `votes` (number of people who rated the movie), using a scatterplot.

```
ggplot(movies, aes(votes, rating)) + geom_point() + ylim(1,10)
```

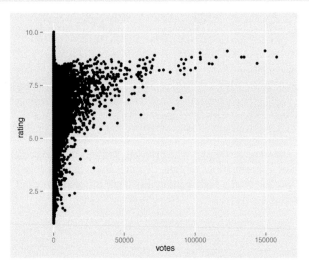

FIGURE 5.7: A scatterplot of the average ratings of films by the number of people who voted, from the dataset *movies*. Films rated very often have higher average ratings, but the highest ratings are achieved by films that are rated far less often.

Figure 5.7 shows the somewhat unexpected result. The scatterplot almost takes the form of a small letter 'r'. There are many insights that can be gained from this plot:

1. There are no films with lots of votes and a low average rating.

2. For films with more than a small number of votes, the average rating increases with the number of votes.

3. No film with lots of votes has an average rating close to the maximum possible. There almost seems to be a barrier, which cannot be crossed.

4. A few films with a high number of votes, over 50,000, look like outliers. They have a distinctly lower rating than other films with similar numbers of votes.

5. Films with a low number of votes may have any average rating from the lowest to the highest.

6. The only films with very high average ratings are films with relatively few votes.

Scatterplots can reveal a lot of information.

5.4   Adding models: lines and smooths

People are good at spotting patterns that are not really there (and are doubtless also good at missing some patterns that really are there). If you think there is a linear causal relationship between the two variables in a scatterplot, it makes sense to fit a model and to add it to the display.

Cars and mpg

The dataset *Cars93* from **MASS** contains 27 pieces of information for 93 cars. The data were collected around twenty years ago [Lock, 1993]. Plotting MPG.city against Weight clearly shows a nonlinear relationship, as fuel economy decreases with weight quite quickly initially and then more slowly (Figure 5.8).

```
data(Cars93, package="MASS")
ggplot(Cars93, aes(Weight, MPG.city)) + geom_point() +
      geom_smooth(colour="green") + ylim(0,50)
```

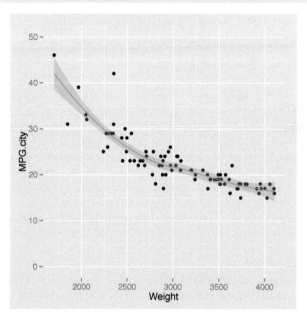

FIGURE 5.8: A scatterplot of MPG.city against Weight for the *Cars93* dataset. A smooth has been overlaid. Heavier cars get progressively fewer miles to the gallon. The car weighing about 2350 pounds with a relatively high mpg was the Honda Civic. The *y* axis has been drawn from 0 to emphasise the lower limit of the mpg value.

Pearson heights

Figure 5.9 shows the fathers and sons height data of Pearson (already discussed in Section 3.3) with the best linear fit added, together with a 95% pointwise confidence interval. The $y - x$ diagonal has been drawn for comparative purposes.

```
data(father.son, package="UsingR")
ggplot(father.son, aes(fheight, sheight)) + geom_point() +
      geom_smooth(method="lm", colour="red") +
      geom_abline(slope=1, intercept=0)
```

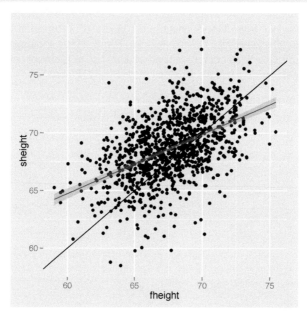

FIGURE 5.9: A scatterplot of sons' heights against fathers' heights from the dataset *father.son*. The best fit regression line has a slope of just over 0.5, as can be seen by comparison with the line $y = x$. The height of a man is influenced by the height of his father, but there is a lot of unexplained variability (the correlation is almost exactly 0.5).

Figure 5.9 illustrates regression to the mean. Tall fathers have sons who are tall, but on average not as tall as their fathers. Small fathers have sons who are small, but on average not as small as their fathers. The fit of the model can be examined with

```
data(father.son, package="UsingR")
m1 <- lm(sheight~fheight, father.son)
summary(m1)
par(mfrow=c(2,2))
plot(m1)
```

To explore further whether a non-linear model might be warranted, you could fit a smoother or plot a smoother and the best fit regression line together. For these data, a linear model is fine, since the two curves are practically identical, as you can see in Figure 5.10.

```
data(father.son, package="UsingR")
ggplot(father.son, aes(fheight, sheight)) + geom_point() +
    geom_smooth(method="lm", colour="red", se=FALSE) +
    stat_smooth()
```

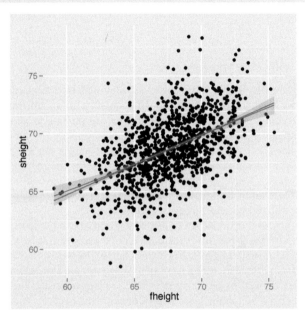

FIGURE 5.10: A scatterplot of sons' heights against fathers' heights with both a linear regression line and a smoother overlaid. The two models are in close agreement and the line lies inside the confidence interval of the smooth. A lot of unexplained variability remains.

Adding a line or a smooth (or both) makes modelling explicit. When you look at a graphic you employ implicit models in judging what you see and it is good to formalise these where possible. With something like heights of fathers and sons you might expect some kind of positive association and judge how the plot looks with that image in mind. In other situations, perhaps when the same plot includes both males and females, you might expect to see signs of there being two groups. Prior expectations can be based on context and existing knowledge. Once you have seen a plot, the data themselves influence what features you study and how you judge them. Subjective impressions can supply valuable insights, but need to be confirmed by checking with other views of the data and with objective testing, where possible.

5.5   Comparing groups within scatterplots

The Olympic athletes' dataset shown in Figure 5.1 suffered from overplotting, which is one good reason for adjusting the scatterplot when displaying the data. Either overlaid density estimates or alpha-blending might be used, but given the structure of the data, where we can expect height and weight to differ by the sex of the athletes and their sport, a better approach would be to split up the data by possible explanatory variables. For instance, the following code plots a scatterplot for the females above one for the males:

```
ggplot(oly12, aes(Height, Weight)) +
      geom_point(size = 1) + facet_wrap(~Sex, ncol=1)
```

There are 42 sports listed in the dataset and plotting all the scatterplots together gives Figure 5.11. Although each plot is quite small, you can still easily identify a number of features. For some sports, all athletes are missing at least one of the values (e.g., boxing, gymnastics) and for others, there are very few athletes. Where there are enough data points, the association between height and weight generally holds well, apart from some outliers. It is noticeable that the relationship is less good for athletics, where a large number of very different events have all been grouped together. With judo and wrestling, the relationship is also weaker, possibly because of the range of different weight classes in these sports.

Whenever you split up a dataset in this way it is important to have all the plots the same size with the same scales for comparative purposes. This is handled automatically with facetting in **ggplot2** or within **lattice**. However, it is also useful to organise the plots, so that the comparisons of most interest are easiest to make. Sometimes having certain plots in the same row is best, sometimes having them in the same column. And often it will make sense to pick out particular subsets and make a plot just with them. If you want to compare the height and weight scatterplots for judo, weightlifting, and wrestling, then create the subset first and plot accordingly, e.g.

```
oly12JWW <- filter(oly12, Sport %in%
            c("Judo", "Weightlifting", "Wrestling"))
ggplot(oly12JWW, aes(Height, Weight)) +
      geom_point(size = 1) + facet_wrap(~Sport) +
      ggtitle("Weight and Height by Sport")
```

```
oly12S <- within(oly12, Sport <- abbreviate(Sport, 12))
ggplot(oly12S, aes(Height, Weight)) +
       geom_point(size = 1) + facet_wrap(~Sport) +
       ggtitle("Weight and Height by Sport")
```

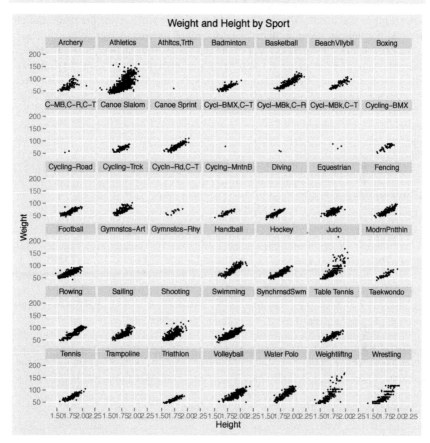

FIGURE 5.11: Scatterplots of weight by height for the different types of sport at the London Olympics. Data are missing for some sports. The association between weight and height looks linear for most sports. The names have been limited to a maximum of 12 characters using the abbreviate function.

5.6   Scatterplot matrices for looking at many pairs of variables

Scatterplot matrices (sploms) are tables of scatterplots with each variable plotted against all of the others. They give excellent initial overviews of the relationships between continuous variables in datasets with small numbers of variables.

There are many different ways of drawing sploms. You can have different options of what is plotted on the diagonal or of what is plotted above or below the diagonal. Since the plots above are just flipped versions of the ones below, some analysts prefer to provide statistical information or other displays in the other half of a splom. You can restrict sploms to continuous variables or provide additional types of plot for including categorical variables. If you plot histograms or density estimates down the diagonal, then the scatterplot matrix gives you an overview of the one-dimensional distributions as well.

Crime in the U.S.

The dataset *crime.us* from the **VGAMdata** package includes the absolute crime figures and the crime rates by population for the fifty U.S. states in 2009. Figure 5.12 shows a splom of the rates for seven kinds of crime (the dataset also includes rates for the four crimes of violence together and the three property crimes together). It also shows density estimates for the variables down the diagonal. Whatever preconceived notions we might have of how the rates might be related to one another, the splom provides a first, quick summary of the associations between the variables, both through the graphics and through the correlation coefficients.

The scales have not been drawn for two reasons. They overlap with the variable names and they are too small to read easily on a graphic this size (larger versions work better if you want to see the scales). It is important to remember that the rate levels are quite different for different crimes. State murder rates in this dataset go up to almost 12 per 100,000 population, whereas larceny rates go up to just over 2700 per 100,000 population.

The rates for most crimes seem positively associated, some more strongly than others (e.g., murder and burglary), and the rate of rape is not associated much with any of the other crime rates. The larceny rate is not closely associated with the four violent crime rates. There are a few outliers, which stand out more in some plots than others, such as the two states with high rates of motor vehicle theft in the scatterplot with larceny (California and Nevada). As the dataset only has 50 cases, you cannot read too much into the shape of the graphics or the correlation coefficients. It is still instructive to compare each plot with its correlation to see how they compare. The different sizes of the State populations should also be taken into account. Alaska, North Dakota, Vermont, and Wyoming are each treated as having equal weight with California, even though their populations are less than 2% of California's.

```
data(crime.us, package="VGAMdata")
crime.usR <- crime.us
names(crime.usR) <- gsub("*Rate", "", names(crime.usR))
names(crime.usR)[19:20] <- c("Larceny", "MotorVTheft")
ggpairs(crime.usR[, c(13:16, 18:20)],
        title="Crime rates in the USA",
        diag=list(continuous='density'), axisLabels='none')
```

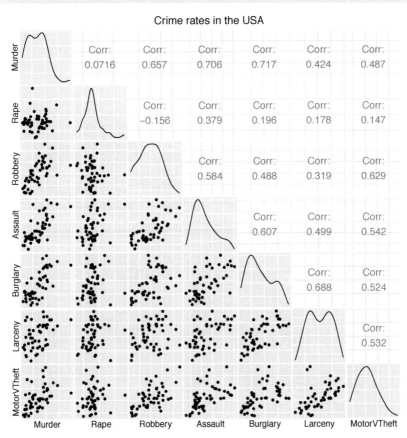

FIGURE 5.12: A scatterplot matrix of the 7 crime rates in the *crime.us* dataset. Some rates are positively associated, others not. The crime of rape is least associated with the others.

Swiss banknotes

The dataset *bank* from the **gclus** package includes six measurements on each of 100 genuine Swiss banknotes and 100 forged notes. The dataset was used extensively in a fine multivariate statistics textbook [Flury and Riedwyl, 1988].

```
library(car)
data(bank, package="gclus")
par(mar=c(1.1, 1.1, 1.1, 1.1))
spm(select(bank, Length:Diagonal), pch=c(16, 16),
    diagonal="histogram", smoother=FALSE,
    reg.line=FALSE, groups=bank$Status)
```

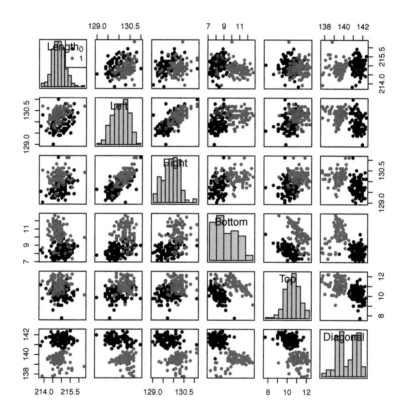

FIGURE 5.13: A scatterplot matrix of the Swiss banknotes dataset with the forged notes coloured in red. In some of the scatterplots the groups of notes are well separated. Some variables are associated, some not. There are a few possible outliers, not all of them forgeries.

A splom of the variables is shown in Figure 5.13. `Left` and `Right` are strongly positively correlated. There is evidence of negative association amongst the last three variables (`Bottom`, `Top`, and `Diagonal`) and there are also suggestions that the genuine and counterfeit notes can be distinguished using those variables. The scatterplot for `Bottom` and `Top` is particularly interesting, as the overall association is slightly positive, while the two possible subgroups each have a negative association between the two variables. The default display for the `spm` function from **car** includes smoothers with confidence bands, linear regression lines in the scatterplots, and density estimates on the diagonal. It also uses open circles for the points. All this makes for a fairly cluttered display, so other options have been chosen here.

Functions for drawing sploms

In R you can choose between the functions listed in Table 5.1 and probably there are other alternatives too. Some of these are faster than others.

function	package
`plot`	
`pairs`	
`spm`	**car**
`cpairs`	**gclus**
`splom`	**lattice**
`ggpairs`	**GGally**
`gpairs`	**gpairs**
`pairs.mod`	**SMPracticals**
`pairscor.fnc`	**languageR**
. . .	. . .

TABLE 5.1: Splom functions in R

Each of the functions offers different versions. In particular, some can cope with categorical variables and some cannot. In this chapter we have looked at some simple scatterplot matrices involving just continuous variables (and only a few of them at that). If you want to look at more elaborate sploms with more variables, take a look at the examples on the help pages of some of the functions, for instance:

• `cpairs` with panels coloured by level of correlation

```
library(gclus)
judge.cor <- cor(USJudgeRatings)
judge.color <- dmat.color(judge.cor)
cpairs(USJudgeRatings, panel.colors=judge.color,
       pch=".", gap=.5)
```

- `gpairs` with loess smooths

```
library(gpairs)
data(Leaves)
gpairs(Leaves[1:10], lower.pars=list(scatter='loess'))
```

- `pairs.mod` with standard scatterplots as well as scatterplots of partial residuals

```
library(SMPracticals)
data(mathmarks)
pairs.mod(mathmarks)
```

Sploms take up a lot of space. A splom for m variables includes $m(m-1)$ scatterplots and m diagonal elements. Even if only the upper triangle of plots is drawn, you still need m rows and m columns for m variables. This makes sploms ineffective for large numbers of variables, especially when there are many cases, as each individual plot can only be drawn fairly small. Nevertheless, for a quick overview, perhaps for deciding which of a number of scatterplots to look into in more detail, they can be very useful.

5.7   Scatterplot options

- **Point size**
  Very small points can hardly be seen and very large points overlap each other and make the plot look like a collection of clumps. Making points a little larger can be useful for emphasising outliers; making points a little smaller can be useful for distinguishing groups of points. Occasionally, particular point sizes can lead to undesirable visual effects, especially if the data are gridded with parallel strips of points close together.

  Point size can also be used to represent the value of a continuous non-negative variable. When the basic symbol is a circle, then this gives a bubble chart. It can be quite useful for smallish datasets without too much overlapping and is used in Gapminder's displays of global development data [Rosling, 2013].

- **Symbols for points**
  The current default in R for scatterplots is to use open circles. As attentive readers will have gathered, I prefer small filled circles. You should use what you think conveys information best for you.

  When a dataset is small and made up of different groups, it is helpful to be able to tell the groups apart in a scatterplot. An old solution is to draw the members of each group with a different symbol. This can work if the points do not overlap and if the symbols are easy to identify. Mostly it does not, and colour is more effective anyway. With large datasets, using different symbols looks cluttered and messy, so a set of scatterplots, one for each group, as in a trellis display, works better.

- **Alpha-blending**
  A partial solution to overplotting problems is to use alpha-blending. Each point is given a weight between 0 and 1 and where several points overlap, the resulting area is drawn correspondingly darker (in fact more opaque). For instance, if $\alpha = 0.1$, then any area with ten or more points has the maximum darkness. The effect is to emphasise areas of higher density and downplay areas of lesser density, so that outliers cannot be seen so easily. Using no alpha-blending (equivalent to $\alpha = 1$) can leave the bulk of the data for a large dataset looking like a solid indistinguishable mass. Alpha-blending works better interactively, when you can explore a range of alpha values quickly to see what information is shown at different levels.

- **Colouring points**
  Colour is often used to distinguish points by groups. If the data are fairly well separated (whatever that means in practice), this is helpful. You always have to bear in mind that when there are points on top of one another, the visible colour is the colour of the last point drawn. So if there are three colours, red, green, and blue, the plot may look different if drawn in that order (maybe lots of blue), than if drawn in the reverse order (maybe lots of red).

  If you think an interesting structure is evident in a coloured scatterplot, then it is worthwhile drawing a trellis plot of scatterplots by group. Although coloured scatterplots may have disadvantages, it is often easier to identify a possible group structure there than in a trellis plot and it is always good to check.

  When colour is used to highlight particular points, it is usually drawn last and hence on top. The same cautionary advice applies here as for colouring by groups.

- **Splom options**
  The many options available for sploms were referred to in §5.6. You can display statistics instead of half of the scatterplots, add models to scatterplots, or show marginal information on the diagonal.

5.8   Modelling and testing for relationships between variables

1. Correlation

   If you calculate a correlation coefficient, you should always draw a scatterplot to learn what the coefficient might mean. Correlation coefficients measure linear association and it is a rare scatterplot where that is all you can see. By the same token, if you draw a scatterplot and see a linear association, it is a good idea to calculate the correlation coefficient to measure just what level of correlation you have in front of you (cf. [Cleveland et al., 1982]).

   Correlation coefficients are sometimes accompanied by p-values in publications and the null hypothesis tested always seems to be $H_0 : \rho = 0$. Fisher showed long ago in 1915 that there is a good approximation for testing null hypotheses of correlation coefficients being equal to more interesting and informative values than 0, but this is seldom done. Pity.

2. Regression

   Given a presumed causal relationship between the y-variable and the x-variable of a scatterplot, regression may be used to fit a model. It can be helpful to overlay the model on a scatterplot of the data, to add confidence bands for the fit, or to add predictive intervals for possible new points.

3. Smoothing

   If no analytic model is proposed for Y as a function of X, then a nonlinear smoother can be tried. `loess` (local weighted regression) is an interesting approach and is used in some situations in R as a default, for instance in plots of model objects. Recently spline functions have been used more, partly because they have better statistical properties and because it is easier to draw confidence bands for them.

4. Bivariate density estimation

   `kde2d` from **MASS**, `kde` from **ks**, or `bkde2d` from **KernSmooth** may be used. The highest density region package **hdrcde** is also an interesting possibility. As with univariate density estimation, there is no testing of the estimates produced, a curious gap in theory.

5. Outliers

   Points which are outliers in scatterplots may be outliers on one of the two dimensions individually or purely bivariate outliers. Deciding whether a point is a bivariate outlier or whether several points in a group are outliers is tricky. Density estimators are one approach, but they can be problematic because of either masking or swamping.

Main points

1. Scatterplots can take many different forms and provide a lot of information about the relationship between two variables (cf. §5.2). Look again at the movies scatterplot in Figure 5.7.

2. Adding lines or smooths to scatterplots is easy and often provides valuable guidance, as can be seen in Figure 5.9, where the relationship between sons' and fathers' heights is difficult to assess from the scatterplot alone.

3. Trellis displays are very effective for comparing scatterplots by subgroup, especially when the groups overlap (cf. Figure 5.11).

4. Splooms are excellent for giving a quick overview of a few variables. For example, Figure 5.12 summarises the U.S. crime rates dataset well.

Exercises

1. **Movie ratings**
   In Figure 5.7, there were a number of films with very high average ratings but few votes.

   (a) How does the scatterplot look if you exclude all films with fewer than 100 votes?
   (b) What about excluding all films with an average rating greater than 9? In the latter case, you would exclude at least one film with a lot of votes. What limit would you choose and why?

2. **Meta analysis**
   The results of 70 studies on thrombolytic therapy after acute myocardial infarction are reported in the *Olkin95* datatset from the **meta** package.

   (a) Draw a scatterplot of the number of observations in each experimental group (n.e) against the corresponding number of observations in each control group (n.c). How would you summarise the plot?
   (b) Over half of the studies involve fewer than 100 patients in each group. If you restrict the scatterplot to this range, do you gain additional information?

3. **Zuni**

   The *zuni* dataset from the package **lawstat** was introduced in Exercise 5 in Chapter 3.

   (a) Plot a scatterplot of average revenue per pupil (Revenue) against the corresponding number of pupils (Mem). What information can you gather from the plot?

   (b) The distribution of the number of pupils is skew with one large outlier. How does the scatterplot look if you plot Revenue against the log of the number of pupils? What additional information have you discovered, if any? Is there any point in logging average revenue per pupil as well?

4. **Pearson heights**

   Pearsons' height data for fathers and sons were considered in Sections 3.3 and 5.4.

   (a) Draw a scatterplot of the heights. Are there any cases you would regard as outliers?

   (b) Draw a plot including both points and highest density regions. Which cases would be regarded as outliers under this model, do you think?

   (c) Fit a linear model to the data and a loess smooth and plot your results. Is a nonlinear model necessary?

5. **Bank discrimination**

   A subset of Roberts' bank sex discrimination dataset from 1979 is available in the package **Sleuth2** under the name *case1202*. Consider the three variables measured in months, Senior (seniority), Age, and Exper (work experience prior to joining the bank).

   (a) Are there any notable features in the scatterplot matrix of the three variables? Can you explain them?

   (b) Why do the scatterplots involving seniority not have the structure of the scatterplot of experience against age?

6. **Cars**

   In Figure 5.8, MPG.city was plotted against Weight. In many countries, fuel performance is measured in litres per 100 km rather than in miles per gallon, in effect the inverse criterion. If you plot 1/MPG.City against Horsepower, do you get a linear relationship? Which cars would you describe as outliers now?

7. **Leaves**

   The *leafshape* dataset in the **DAAG** package includes three measurements on each leaf (length, width, petiole) and the logarithms of the three measurements.

   (a) Draw sploms for the two sets of three variables. What conclusions would you draw from each set? Which do you find more useful?

   (b) Redraw the sploms, colouring the cases by the variable arch, describing the type of leaf architecture. What additional structure can you see?

8. **Olive oils from Italy**
   The olive oils dataset is well known and can be found in several packages, for instance as *olives* in **extracat**. The original source for the data is the paper [Forina et al., 1983].

   (a) Draw a scatterplot matrix of the eight continuous variables. Which of the fatty acids are strongly positively associated and which strongly negatively associated?
   (b) Are there outliers or other features worth mentioning?

9. **Boston housing**
   The Boston dataset was introduced in Chapter 3.

   (a) Draw a splom of all the continuous variables (i.e., all except the variable chas). Which variables are positively associated with medv, the median home value?
   (b) Several of the scatterplots involving the variable crim, the per capita crime rate, have an unusual form, where higher values of crim only occur for one particular value of the other variable. How would you explain this?
   (c) There are many different scatterplot forms in the display. Pick out five and describe how you would interpret them.

10. **Hertzsprung-Russell**
    The Hertzsprung-Russell diagram is a famous scatterplot of the relationship between the absolute magnitudes of stars and their effective temperatures and is over one hundred years old. Although examples of the plot can be found all over the place, it is surprisingly difficult to find the data underlying them. There is a dataset of 47 cases, *starsCYG*, in the package **robustbase**, but that is really too small. The dataset *HRstars* with 6220 stars in package **GDAdata** is from the Yale Trigonometric Parallax Dataset and was downloaded from [Mihos, 2005].

    (a) Plot Y against X. How does your plot differ from the plots you find on the web, for instance from a Google search for images of the Hertzsprung-Russell diagram?
    (b) The plots seem to use different numbers of stars. Are some more likely to be used than others?
    (c) You can colour and annotate your plot using techniques described in Chapter 13. What would you suggest?

11. **Intermission**
    The painting *La Grande Jatte* by Georges Seurat hangs in the *Art Insitute of Chicago*, a classic of putting dots together to form an overall impression. Do you think the artist intended his painting only to be viewed from a distance?

6

Investigating Multivariate Continuous Data

Numerical quantities focus on expected values, graphical summaries on un-expected values.

John Tukey

Summary

Chapter 6 discusses parallel coordinate plots for studying many continuous variables simultaneously.

6.1   Introduction

Scatterplots are superb for looking at bivariate data, but they are not so effective for exploring in higher dimensions. Matrices of scatterplots are pretty good, but do not really convey what multivariate structure there might be in the data, even with inter-active linking. Dimension reduction methods (principal component analysis, factor analysis, or multidimensional scaling) have their supporters, and like all graphics-related approaches can provide insights when they are appropriate for the dataset in hand. Their disadvantage is that they approximate the data, and it is difficult to assess how good the approximations are. Rotating plots, which dynamically rotate through two-dimensional projections, are a particularly flexible dimension reduction approach. They are attractive and can uncover interesting information when com-bined with projection pursuit indices, but are really a specialist tool.

In recent years, parallel coordinate plots have become popular for multivariate continuous data. They were discussed in depth by Inselberg, and their geometry is covered in detail in his book [Inselberg, 2009]. Wegman was the first to suggest their use for data analysis [Wegman, 1990]. Chapter 1 of [Hartigan, 1975] talks about pro-file plots, a very similar idea, which was inhibited in practice at the time by the com-puting power available. It is interesting to look at Hartigan's plots and be impressed both by what he achieved then and by how far graphics have come since.

6.2   What is a parallel coordinate plot (pcp)?

```
data(food, package="MMST")
names(food) <- c("Fat", "Food.energy", "Carbohyd", "Protein",
                 "Cholest", "Wt", "Satur.Fat")
ggparcoord(data = food, columns = c(1:7), scale="uniminmax") +
           xlab("") + ylab("")
```

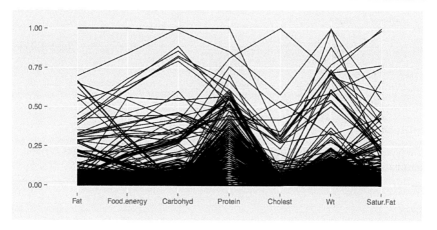

FIGURE 6.1: A parallel coordinate plot of the seven variables in the *food* dataset from the **MMST** package. There are outliers and all the distributions look skewed to the right, because most of the lines are packed into the foot of the plot. Setting `scale="uniminmax"` means that each variable is scaled individually from its minimum to its maximum.

With a scatterplot, the *x* and *y* axes are perpendicular to one another. In a parallel coordinate plot all axes are parallel to one another. Each variable has its own individual vertical axis (or alternatively all the axes are horizontal) and the axis is usually scaled from the minimum to the maximum case values for the variable, so that the full range of each axis is used. The values of each case on adjacent axes are joined by lines, so that a polygonal line across all axes defines a case. Hartigan called these lines 'profiles', and that is a good way of thinking of them. Figure 6.1 shows a pcp of the seven variables of the *food* dataset from the **MMST** package.

The *food* dataset has 961 cases. The variables all look skewed to the right, as there are more lines at the bottom of the plot than at the top. One case is an extreme value on several of the variables (the line starting top left). There appears to be a small subgroup with similar sets of highish values (the group of lines close together across the first five variables).

Izenman's textbook [Izenman and Sommer, 1988] suggests dividing each food value by the serving weight of the food and Figure 6.2 shows the corresponding parallel coordinate plot (without the weight variable, of course). There are now a couple of distinct groups at high levels of the fat by weight variable. The protein by weight and the cholesterol by weight variables each have an extreme outlier.

```
food1 <- food/food$Wt
ggparcoord(data = food1, columns=c(1:5, 7),
           scale="uniminmax", alphaLines=0.2) +
           xlab("") + ylab("")
```

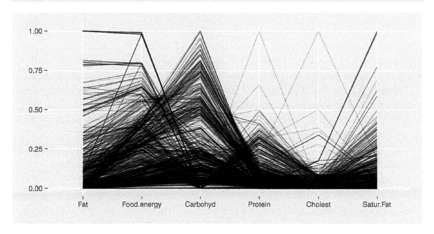

FIGURE 6.2: A parallel coordinate plot of the six variables in the *food* dataset divided by the variable `weight.grams`, the weight of a serving. The transformed carbohydrates variable is no longer skew and there are some distinct groups with high fat levels. Individual outliers can be seen on cholesterol and protein and a small group on saturated fat. The alpha-blending parameter `alphaLines` (§6.7) has been used to make the plot easier to read. Each case is plotted with a weight of 0.2, so that fully black lines represent at least 5 cases.

Parallel coordinate plots can include axes for categorical variables as well, where all cases for a particular category are assigned the same numerical value. It is advisable to avoid drawing the axes for categorical variables beside one another, as then many lines are drawn on top of one another and it is difficult to see structure.

Functions for drawing pcp's

There are several functions for pcp's in R and `ggparcoord` is the one primarily used in this chapter. It has a lot of options, both for what is displayed and how it is displayed, which is a good thing, because there are a lot of possible ways of drawing pcp's and these can have a big impact on what you see. As always, different people have different preferences and it is sensible to experiment a little to find out what works best for you. Just settling for the defaults is a flawed strategy, particularly

function	package
ggparcoord	**GGally**
parcoord	**MASS**
cparcoord	**gclus**
pcp	**PairViz**
parallelplot	**lattice**
parallel.ade	**epade**
ipcp	**iplots**
. . .	**. . .**

TABLE 6.1: pcp functions in R

when there are so many alternatives offered, but you do have to put in the work to find out what the options do and how the resulting displays should be interpreted.

Other functions listed in Table 6.1 include parcoord, which offers basic functionality including colouring, and cparcoord in **gclus** is a modified version of it, as is pcp. The function parallel.ade offers a scaling alternative and so does ipcp in **iplots**, which has the additional advantage of being interactive.

In fact, parallel coordinate plots really need to be interactive to be fully effective. Highlighting cases which are of interest on one axis and seeing where they appear on the other axes instantaneously is a simple and intuitive process. Specifying an appropriate condition to define the group of interest and redrawing the plot each time is laborious. Reordering axes on the fly is more effective than writing out the necessary function and drawing the plot again. A good strategy is to use a tool like ipcp initially to ascertain what information is of interest and how you want to display it, and then draw the chosen plot using your favourite static pcp function.

6.3   Features you can see with parallel coordinate plots

[Wegman, 1990] suggested that it is possible to identify many different multivariate features in parallel coordinate plots. This can be true in particular applications, but in general the claim is too optimistic. What is definitely true is that having identified outliers, correlations, or clusters with analytic approaches, parallel coordinate plots are very useful for checking the results.

Parallel coordinate plots give quick overviews of univariate distributions for several variables at once: whether they are skew, whether there are outliers, whether there are gaps or concentrations of data (all of these can be seen in Figures 6.1 and 6.2). Bivariate associations between adjacent variables can sometimes be seen and occasionally some multivariate structures, such as groups of cases with very similar values across a number of variables or cases which are outliers on more than one variable (as in Figure 6.1). These patterns are best checked and investigated further by colouring the group of interest. Figure 6.3 shows an example for the *food* dataset, drawn horizontally this time using `coord_flip`. Note that the cases have been re-ordered to ensure that the selected cases are drawn last, so that those lines are on top.

```
food1 <- within(food1,
                fatX <- factor(ifelse(Fat > 0.75, 1, 0)))
ggparcoord(data = food1[order(food1$fatX),],
           columns=c(1:5, 7), groupColumn="fatX",
           scale="uniminmax") + xlab("") + ylab("")  +
           theme(legend.position = "none") + coord_flip()
```

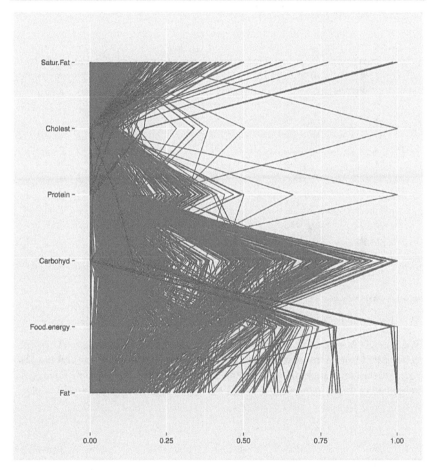

FIGURE 6.3: A parallel coordinate plot of the six variables in the *food* dataset divided by serving weight. The two subgroups with high values of fat per serving have been selected using the derived variable `fatX`. As expected, some of these cases have the highest values for saturated fat. This is another version of Figure 6.2.

Wegman and others have suggested riffling through sufficient reorder-
ings of the axes to observe all possible adjacencies. In a similar vein,
[Hurley and Oldford, 2010] proposed adding extra copies of axes to include all the
adjacencies. If your main aim is to look for bivariate associations, then you are better
off with scatterplots. There is an example in Figure 6.2 where it looks as if the transformed carbohydrates and protein variables may be negatively correlated because of
the lines crossing to form an 'X' shape. In fact the correlation is only -0.087, although the relationship between the variables is worth looking at for other reasons,
as Figure 6.4 shows.

```
ggplot(food1, aes(Protein, Carbohyd)) + geom_point()
```

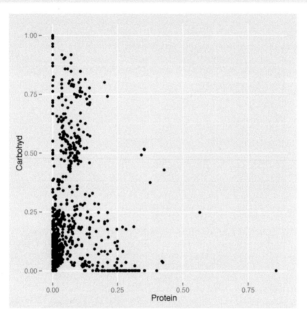

FIGURE 6.4: A scatterplot of carbohydrate and protein values standardised by serving weight. Most of the points have low protein values, and there are quite a few
foods with either no protein or no carbohydrates. The apparent diagonal boundary is
determined by foods composed of almost only protein and carbohydrates.

Fisher's *iris* dataset has been used so often as an example for so many statistical approaches, including in this book, because it makes for such an excellent illustration. In a pcp of the data in Figure 6.5 you can see how clearly the `setosa` species is separated from the other two species on the petal measurements, how it is lower on three of the measurements, but higher on `Sepal.Width`, where there is also a `setosa` outlier, how the two petal measurements almost separate the other two species, and how additionally there are two unusual `virginica` values on `Sepal.Width`.

```
ggparcoord(iris, columns=1:4, groupColumn="Species")
```

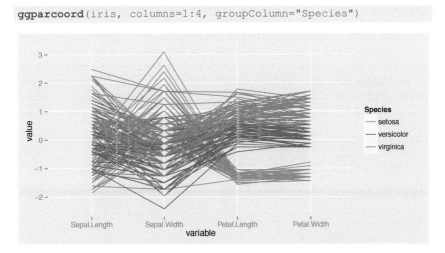

FIGURE 6.5: A parallel coordinate plot of the four variables in the *iris* dataset coloured by species and using the `ggparcoord` default scaling, standardising by the mean and standard deviation. The petal measurements separate the groups well and there is evidence of cases which are a little different from the rest of their species.

There is no guarantee that a pcp will reveal interesting information about a dataset or indeed any information. That applies to all graphics. What can be said is that when graphics can be drawn so quickly and easily it is worth checking them out to see what information they do reveal.

6.4   Interpreting clustering results

Cluster analysis is a popular data analysis tool, although most of the methods have little statistical basis and results are judged more by what users want to do with them than by any statistical approach. Even if some well-separated clusters are obtained, you still need to find a way of describing them and that means working out for which variables or groups of variables they differ from one another and how. A productive first step is to compare the clusters graphically with a parallel coordinate plot in which the clusters are given different colours.

Figure 6.6 is a pcp of the *USArrests* dataset from 1973, coloured by the three clusters found with an average link clustering, the first example on the `hclust` help page. The `Assault` variable separates the clusters, and the reason becomes obvious if you draw the same plot with the option `scale="globalminmax"`, which puts all the variables on a common scale: `Assault` has a much higher range of values than the other three variables, so it dominates the multivariate distance calculations.

```
hcav <- hclust(dist(USArrests), method="ave")
clu3 <- cutree(hcav, k=3)
clus <- factor(clu3)
usa1 <- cbind(USArrests, clus)
ggparcoord(usa1, columns=1:4, groupColumn="clus",
           scale="uniminmax", mapping = aes(size = 1)) +
           xlab("") + ylab("") +
           theme(legend.position = "none")
```

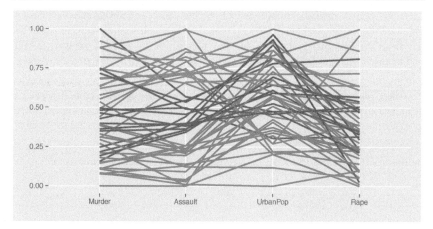

FIGURE 6.6: A parallel coordinate plot of the *USArrests* dataset. The three clusters found with an average linkage clustering have been coloured. The clusters are defined by the `Assault` variable, as can be seen by the clear separation of the three colours on that variable. The `size` parameter has been used to make the lines thicker.

A clustering using the variables more equally could be carried out using `dist(scale(USArrests))` in `hclust`, as Figure 6.7 shows. The `Assault` variable still separates the clusters fairly clearly on its own, but now there are only two main clusters, not three, as one cluster is a state on its own. Because the dataset is small, you can list the cluster members or draw a dendrogram with `plot(hcav2)` to identify the outlying state as Alaska.

```
hcav2 <- hclust(dist(scale(USArrests)), method="ave")
clu32 <- cutree(hcav2, k=3)
clus2 <- factor(clu32)
usa2 <- cbind(USArrests, clus2)
ggparcoord(usa2, columns=1:4, groupColumn="clus2",
           scale="uniminmax", mapping = aes(size = 1)) +
           xlab("") + ylab("") +
           theme(legend.position = "none")
```

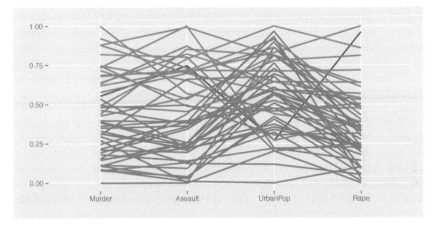

FIGURE 6.7: A parallel coordinate plot of the *USArrests* dataset drawn using scaled variables and average linkage clustering. The three cluster solution is now more of a two cluster solution with one outlying case (which turned out to be Alaska).

Most applications in practice are not this simple and there is an extensive range of additional clustering approaches as well as alternative distance measures that might be used instead. The key point to remember is that whatever clustering method you use it is valuable to explore the results you obtain graphically rather than just accepting them without any further investigation.

6.5   Parallel coordinate plots and time series

Time series are a special kind of data and demand special treatment. There is plenty of support in R for analysing and displaying time series, outlined in the relevant Task View, TimeSeries [Hyndman, 2013]. Interestingly, some time series can also be plotted as pcp's. Since this sheds light on both pcp's and time series, it is worth looking at briefly.

If data are recorded at regular intervals, multiple series can be displayed simultaneously in a pcp with each axis being one of the time points. Even if the data are recorded irregularly, a pcp can still be used, provided that all series have been recorded at the same times, although you have to bear in mind the distortion of the time axis when interpreting any resulting patterns. As data are usually recorded every day, every month, or every year, this is not often an issue. What is an issue is that the time points have to be in the columns and the series in the rows, which either requires a transposition of the data, or, if there are different sets of series, some data reorganisation. This can be seen in the code needed for Figure 6.8, where **reshape**'s melting and casting have been used. Note that it is essential in plotting time series to use a common scaling for all axes (hence the choice of the `scale` option) and that you cannot change the order of the axes because the time ordering is given. In the `parcoord` default implementation any series with any missing values are completely excluded.

Figure 6.8 shows an example of acres of corn planted in states of the United States over the past 150 years or so. The sharp rise for five states in the 19th century, the dips at the time of the Depression in the 1930s and in the early 1980s, and various unusual individual patterns are all visible. Identifying the individual states is not so easy, using an interactive tool like `ipcp` in *iplots* is better for that. It turns out that Kansas was one of the early 'big' states but declined erratically; Nebraska had the steepest fall, but also some large increases; Iowa and Illinois were always 'big'.

Colours were assigned by default alphabetically, which is why Iowa and Illinois look alike. No legend was included to allow more space for the display. This is typical for initial exploratory graphics, first check to see if there is anything worth studying in more detail, then invest time in improving the plot.

Plotting many time series together gives a rapid overview, highlighting general trends (if there are any) and picking out series which follow different patterns to the rest. As always there is no guarantee that a particular graphic will provide immediate information, perhaps there are no notable features to be found. What can be guaranteed is that being able to draw a range of plots quickly is an effective way of gaining first insights into a dataset.

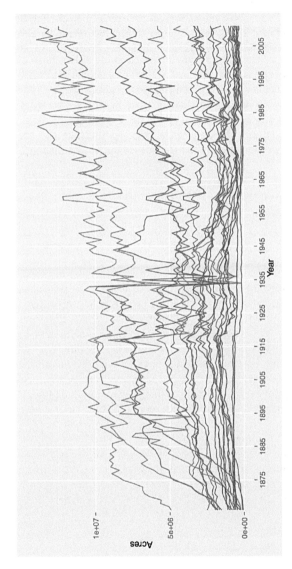

FIGURE 6.8: A pcp of the acres of corn planted by US states from 1866 to 2011 from the *nass.corn* dataset in the **agridat** package. Only the states with more than 250,000 acres planted in 2011 and no missing values are included. Setting scale="globalminmax" means the same scaling is used for all variables. Several interesting patterns can be seen, as discussed in the text.

```
library(reshape2); data(nass.corn, package="agridat")
c1 <- melt(nass.corn, id=c("year", "state"))
c1 <- within(c1, StateV <- interaction(state, variable)); c2 <- dcast(c1, StateV~year)
ggparcoord(subset(c2[1:48,], c2[1:48,147]> 250000), columns=2:147, groupColumn="StateV",
    scale="globalminmax") + xlab("Year") + ylab("Acres") +
    scale_x_discrete(breaks=seq(1865, 2015, 10)) + theme(legend.position = "none")
```

A possible advantage of the pcp view is that it is trivial to switch to a cross-sectional view. Have a look at Figure 6.9, where the only changes to the `ggparcoord` function for Figure 6.8 are the addition of boxplots and the use of alpha-blending to downplay the lines. It is now possible to see that two of the states were individual high outliers almost every year. The long-term trend for the states as a group, represented by the median, can be seen, which is not readily possible in Figure 6.8. The years where many states planted lower numbers of acres are also easier to pick out. Cross-sectional views are useful when we are interested in the distributions of values at particular time points and enable us to make rough judgements as to when individual series may be considered as outliers compared to the rest or not.

The cross-sectional view is also useful for aligning series in different ways (cf. §6.7). Parallel coordinate plots may not be the best option for displaying time series, but they are often a pretty good one. The time series examples illustrate how the number of axes can be very high and yet pcp's can still convey useful information.

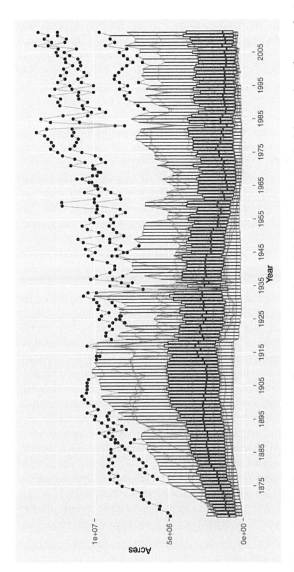

FIGURE 6.9: A parallel boxplot of the *nass.corn* data as specified Figure 6.8. The median trend of the number of acres planted by state can be seen. It is also observable that two states (Iowa and Illinois) are outliers almost every year.

```
ggparcoord(subset(c2[1:48,], c2[1:48,147]> 250000), columns=2:147, groupColumn="StateV",
  scale="globalminmax", boxplot=TRUE, alphaLines=0.5) + xlab("Year") + ylab("Acres") +
  scale_x_discrete(breaks=seq(1865, 2015, 10)) + theme(legend.position = "none")
```

6.6   Parallel coordinate plots for indices

Index values are weighted combinations of the values of their components. Stock market indices are used to represent performances of groups of shares and the consumer price index summarises prices across a wide range of products. Both of these indices are time series, and an obvious graphic display is to overlay the index series in bold on top of the series for the separate components. If a pcp is used for this, then the axes are time points and the cases are the components. Alternatively, indices can be constructed in the same way for different individuals and these individuals can be compared by the single index value and additionally by the component values for each one. In this situation a pcp can be drawn with one axis for the index itself and separate axes for each of the components. The cases are then the individuals and the polygonal lines are their profiles.

The dataset *uniranks* in the package **GDAdata** is an example of this kind of data. 120 universities in the UK were ranked using a combination of eight criteria [Guardian, 2013]. The `AverageTeachingScore` is the overall index and the criteria are the components. Figure 6.10 is a pcp of the data with the variable `StudentStaffRatio` inverted to a staff:student ratio, so that higher values mean a higher overall score for the university. The vertical axis scale has been removed, since it just goes from the minimum to the maximum for each variable, and there might be a temptation to interpret a $(0, 1)$ scale.

A number of features can be seen. High values on the first axis, the index, are mostly associated with high values on the other variables, although not necessarily strongly, and there are clear exceptions (for instance, the third-ranked university has a relatively low value for `NSSTeaching`). The gaps for the three NSS variables reflect limited numbers of possible values for those variables. The up and down pattern for the last few variables is probably due to their being skewed in different directions rather than to any negative association. There are several cases which seem extreme compared to the others on single variables, and there appear to be some bivariate outliers as well, where individual lines go against the trend between two axes.

Plots of this kind are most effective when some individuals or groups are identified by colour. This works best interactively, when selections can be made directly on the graphic and redrawing occurs immediately. With static graphics you need to define a variable to specify the selection you want. If the selection is data-driven, for instance depending on what you see in a first plot, then coding is necessary. If you are interested in predefined groups it is easier and one option could be a trellis display of pcp's. Figure 6.11 shows an example where one of the groups of universities has been picked out.

FIGURE 6.10: A pcp of the *uniranks* dataset from the **GDAdata** package. Three poorly performing universities had at least one missing value on one of the variables and have been excluded. There are a number of outliers and some of the measures are skew.

```
data(uniranks, package="GDAdata")
names(uniranks)[c(5, 6, 8, 10, 11, 13)] <- c("AvTeach", "NSSTeach", "SpendperSt", "Careers",
                                              "VAddScore", "NSSFeedb")

uniranks1 <- within(uniranks, StaffStu <- 1/(StudentStaffRatio))
ggparcoord(uniranks1, columns=c(5:8, 10:14), scale="uniminmax", alphaLines=1/3) +
    xlab("") + ylab("") + theme(axis.ticks.y = element_blank(), axis.text.y = element_blank())
```

The Russell group is made up of 24 of the top UK universities. In Figure 6.11 they have been coloured red to highlight them. The order of the axes has also been changed, based on what the variables represent, so that it now starts with `EntryTarif` and ends with `CareerProspects`. On all variables bar one, `NSSFeedback`, the Russell group have good scores. There is one high-ranking university (St. Andrews, 4th), which is not a member of the Russell group and one, Exeter, with a relatively poor staff:student ratio.

```
uniranks2 <- within(uniranks1,
        Rus <- ifelse(UniGroup=="Russell", "Russell", "not"))
ggparcoord(uniranks2[order(uniranks2$Rus, decreasing=TRUE),],
        columns=c(5:8, 10:14),
        order=c(5,12,8,9,14,6,13,7,11,10),
        groupColumn="Rus", scale="uniminmax") +
        xlab("") + ylab("") +
        theme(legend.position = "none",
        axis.ticks.y = element_blank(),
        axis.text.y = element_blank()) +
        scale_colour_manual(values = c("red","grey"))
```

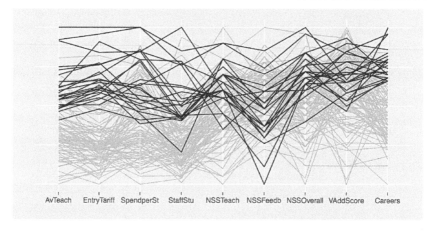

FIGURE 6.11: The *uniranks* dataset with the Russell group selected in red. These universities scored well in general, although not on `NSSFeedback`. The code shows that a new variable specifying the Russell group universities was derived and then used to sort the cases (the first `order`). The order of the variables (columns) was specified by the second `order`.

6.7 Options for parallel coordinate plots

When the pcp is interactive, so that cases can be selected and highlighted, then more information can be uncovered. With static graphics it is necessary to use drawing options in various combinations to get at the information.

Alignment

Instead of aligning the variables by their minimum and maximum values, they can be aligned by their mean or median with a corresponding adjustment to the minimum and maximum limits. This permits comparing the variability across variables better. A particularly instructive example arises in plotting the detailed results from the Tour de France cycle race. A selection of parallel coordinate plots has been used for this on the 'Statistical graphics and more' blog ([Theus, 2013]) for a number of years to present the Tour de France results.

The median time for any stage will represent the time taken by the main bunch of riders, the peloton. This gives a sensible standard against which to judge each rider's performance. On the other hand, it is also informative to align each stage to the time taken by the overall winner of the Tour, as it is immediately apparent on which stages other riders gained time on him. `ggparcoord` provides an option for specifying alignment, although only in conjunction with individual scaling of the axes (`scale="center"` or `"centerObs"`). This means that for a comparison of variabilities on a common scale, you have to transform the data first by subtracting the centering statistic.

The following code shows an example for the *nass.corn* dataset from §6.5 aligning to the `mean`:

```
mz  <- as.data.frame(apply(c2[1:48,2:147], 2,
                       function(x)  x - mean(x,na.rm=TRUE)))
StateV <- c2[1:48,1]
mzA  <- as.data.frame(cbind(StateV, mz))
ggparcoord(mzA, columns=2:147, scale="globalminmax",
           groupColumn="StateV") +
           xlab("Year") + ylab("Acres") +
           scale_x_discrete(breaks=seq(1865, 2015, 10)) +
           theme(legend.position = "none")
```

Scaling

Whenever you want to compare measurements of different kinds you have to find some way of putting them on an equivalent scale. This also arises with other multivariate displays (as will be discussed in §8.3) and with time series (§11.3). There is a surprising number of alternative standardisations you can use and it is as well to think carefully about which one is most appropriate. At any rate, you should always make clear which one you have used, if you present displays to others, and you should equally always check what has been used when interpreting graphics prepared by someone else.

The traditional default scaling for pcp axes depends on the respective minimum and maximum values for each variable. In effect, the vertical scale is drawn from 0 (minimum) to 1 (maximum). (Note that this is not the default scaling used by ggparcoord, which uses the standardised scaling mentioned below.) The basic transformation for case i of variable j is

$$y_{ij} = \frac{x_{ij} - \min_i x_{ij}}{\max_i x_{ij} - \min_i x_{ij}}$$

It is not obligatory to scale the variables individually, and if all variables have equivalent measurement scales it may make sense to put them on a common scale. This would apply if you had exam marks for students for several different subjects. You could then scale all the axes from 0 to 100 (if those were the possible minima and maxima) or from a suitable rounded figure below the minimum across all exams to a rounded figure above the maximum across all exams.

Another approach is to standardise each variable, using either the standard deviation or the IQR (interquartile range), a robust measure of variability. For instance,

$$z_{ij} = \frac{x_{ij} - \bar{x}_j}{sd(x_j)}$$

Figure 6.12 displays four different scalings for the *body* dataset from the **gclus** package. (The variable names have been abbreviated to prevent text overlapping.) Each plot emphasises different aspects and each has a correspondingly different vertical scale.

In the top left panel the outliers on individual variables are picked out. In the top right panel the outliers are still visible and you can see different distributional shapes. In the lower left panel the different value ranges of the variables are emphasised and you get some idea of what scales there are and which variables have similar scales. The display in the lower right panel suggests some possible clusters and offers a different view of the outliers.

```
data(body, package="gclus")
body1 <- body
names(body1) <- abbreviate(names(body), 2)
names(body1)[c(4:5, 11:13, 19:21)] <-
      c("CDp", "CD", "Ch", "Ws", "Ab", "Cl", "An", "Wr")
a1 <- ggparcoord(body1, columns=1:24, alphaLines=0.1) +
               xlab("") + ylab("")
a2 <- ggparcoord(body1, columns=1:24, scale="uniminmax",
               alphaLines=0.1) + xlab("") + ylab("")
a3 <- ggparcoord(body1, columns=1:24,
               scale="globalminmax", alphaLines=0.1) +
               xlab("") + ylab("")
a4 <- ggparcoord(body1, columns=1:24, scale="center",
               scaleSummary="median", alphaLines=0.1) +
               xlab("") + ylab("")
grid.arrange(a1, a2, a3, a4)
```

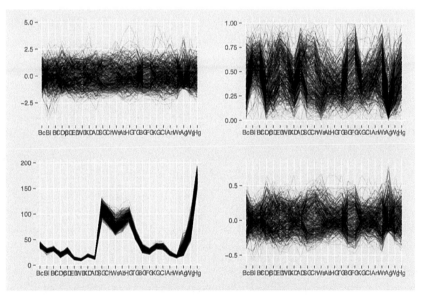

FIGURE 6.12: Four pcp's of the 24 measurements for the 507 men and women in the *body* dataset, each with a different scaling. Top left, the data for each variable have been standardised; top right, each variable has been transformed to a min-max scale individually; lower left, the variables have all been transformed to the same min-max scale; and in the lower right panel the variables have been standardised individually and centred on their medians. Interpretations are discussed in the text.

Outliers

If a variable has outliers, then the rest of the data has to be squeezed into a small section of the variable's axis. Graphics are not robust against univariate outliers. If a form of common scaling is used for a pcp, then this effect is even more restrictive, as one variable with outliers affects the scale for all. While a pcp of the raw data emphasises the outliers, you may want to redraw the plot without them. The code for three alternatives is given below: removing the cases with any outliers, trimming all outlier values to the chosen limits, restricting the plot to the chosen limits.

This code uses the outer fences of boxplots to define outliers, which means they have to be further away from the hinges than three times the IQR. You could use any limits you prefer (and can justify). In each case a function is written to transform the data and applied to the variables to be plotted. Only the first display is shown.

- Remove the cases with outliers (Figure 6.13)
  With 7 variables from the *food* dataset, this resulted in 260 of the 961 cases being excluded, reflecting just how skew the data are. Without the outliers, you can observe regular gaps on the axes for both the fat and protein variables, suggesting a limited discrete range of possible values. Given the large number of cases excluded, it would be unwise to draw too many conclusions. This approach works better when there are only a few outliers to be excluded.

```
fc <- function(xv) {
    bu <- boxplot(xv, plot=FALSE)$stats[5]
    cxv <- ifelse(xv > bu, NA, xv)
    bl <- boxplot(xv, plot=FALSE)$stats[1]
    cxv <- ifelse(cxv < bl, NA, cxv)}
data(food, package="MMST")
rxfood <- as.data.frame(apply(food,2,fc))
ggparcoord(data = rxfood, columns = c(1:7),
           scale="uniminmax", missing="exclude",
           alphaLines=0.3) + xlab("") + ylab("")
```

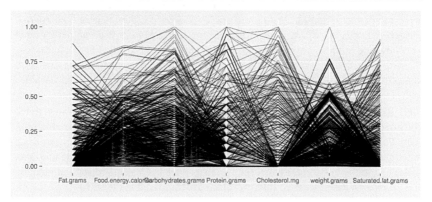

FIGURE 6.13: The *food* dataset with outliers on individual variables removed.

- Trim all outliers to the chosen limits

```
fb <- function(xv) {
    bu <- boxplot(xv, plot=FALSE)$stats[5]
    rxv <- ifelse(xv > bu, bu, xv)
    bl <- boxplot(xv, plot=FALSE)$stats[1]
    rxv <- ifelse(rxv < bl, bl, rxv)}
data(food, package="MMST")
rfood <- as.data.frame(apply(food,2,fb))
ggparcoord(data = rfood, columns = c(1:7),
            scale="uniminmax", alphaLines=0.3)
```

This plot includes all the cases, but 260 of them have been trimmed to the outer hinges of at least one of the variables. It is top-heavy at the outer limits when there are lots of outliers and all the outliers at each end of a variable are plotted together.

- Restrict the plot to the chosen limits

```
fd <- function(xv) {
    bu <- boxplot(xv, plot=FALSE)$stats[5]
    bl <- boxplot(xv, plot=FALSE)$stats[1]
    dxv <- (xv - bl)/(bu - bl)}
data(food, package="MMST")
rofood <- as.data.frame(apply(food,2,fd))
ggparcoord(data = rofood, columns = c(1:7)) +
            coord_cartesian(ylim=c(0,1))
```

Lines going off the top and bottom of the plot are connections to outliers outside the range of the plot. For this dataset, with so many outliers, it is not an effective display.

Each of the three approaches defines cases as outliers, which are outlying on at least one of the variables. The more variables there are, the larger the number of cases which may be regarded as outliers. Cases which are multivariate outliers, but not univariate outliers, are not excluded, because they do not affect the plot scales.

As always when dealing with outliers, these are subjective decisions that need to be considered with care. A further alternative would be to transform the variables with outliers. Sometimes a log transformation can be helpful, although you have to watch for negative values and there may still be outliers even after a transformation. You can see this in Figure 6.2, where an attempt was made to standardise the variables using the serving weight variable. Note that a linear transformation of a variable has no effect on its display in a default pcp, but could affect its display if outliers are taken account of in some way.

Variable order

Any multivariate display is affected by the order of the variables, and default alphabetic orders can be less than informative. In Figure 6.11, the order was manually specified to better match the progression of students through the university. `ggparcoord` has an option, `order`, offering a variety of data-driven alternatives, all of which use scale-free statistics. One approach is to order the axes by the *F* statistics from analyses of variance based on a grouping variable.

```
body1$Gn <- factor(body1$Gn)
ggparcoord(body1, columns=1:24, scale="uniminmax",
           alphaLines=0.4, groupColumn="Gn",
           order="allClass") + xlab("") + ylab("") +
           theme(legend.position = "none",
           axis.ticks.y = element_blank(),
           axis.text.y = element_blank())
```

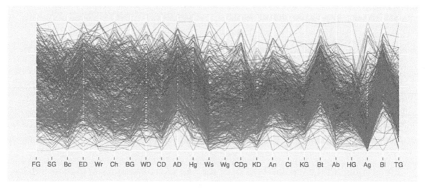

FIGURE 6.14: The *body* dataset with variable axes ordered by differences between the genders as measured by *F* statistics. Women are in red. Forearm and shoulder measurements have the highest *F* values.

In Figure 6.14 the cases of the *body* dataset have been ordered and coloured by gender, while the variables have been ordered by *F* statistics. This static image shows that the variables on the left discriminate clearly between men and women (the men always having higher values) and that there are some outliers, such as the woman with very high values on four variables in a row or the man with the smallest measurement on An (ankle girth). It suffers badly from overplotting for the variables where the genders overlap. Two graphics would be better, one for women with men in grey in the background and one for men with women in grey in the background. A trellis plot would draw only the men in one plot and only the women in the other. Plotting the other gender in the background maintains context and simplifies keeping the scales the same. Figure 6.15 shows the results. It is necessary to order the cases to have the selected group plotted on top. The colours have been chosen to agree with the colours used in Figure 6.14.

```
a <- ggparcoord(body1[order(body1$Gn),], columns=c(1:24),
          groupColumn="Gn", order="allClass",
          scale="uniminmax")  + xlab("") + ylab("") +
          theme(legend.position = "none",
          axis.ticks.y = element_blank(),
          axis.text.y = element_blank()) +
          scale_colour_manual(values = c("grey","#00BFC4"))
b <- ggparcoord(body1[order(body1$Gn, decreasing=TRUE),],
          columns=c(1:24), groupColumn="Gn", order="allClass",
          scale="uniminmax")  + xlab("") + ylab("") +
          theme(legend.position = "none",
          axis.ticks.y = element_blank(),
          axis.text.y = element_blank()) +
          scale_colour_manual(values = c("#F8766D","grey"))
grid.arrange(a,b)
```

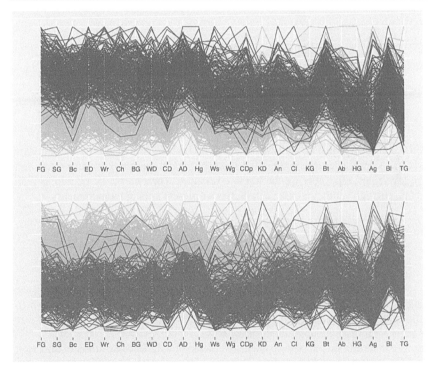

FIGURE 6.15: pcp's of the *body* dataset for men (above) and women (below) with the cases for the other gender drawn in grey in the background. Compare with Figure 6.14, where overplotting is an issue, particularly for the variables to the right. Outliers by sex are now much more apparent.

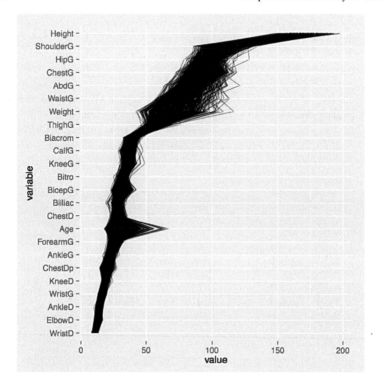

FIGURE 6.16: A pcp of the *body* dataset with the variables sorted by their median values. All variables are on the same scale. In general the variability increases with higher values, especially for the top few variables. The variable in the middle with relatively high variability is Age (which is a different kind of measurement).

```
m2 <- apply(body[, 1:24], 2, median, na.rm=TRUE)
m2a <- order(m2)
ggparcoord(data = select(body, -Gender), alphaLines=0.3,
    scale="globalminmax", order=m2a) + coord_flip()
```

If you want to sort the variables by a statistic, such as the mean, the IQR, or the maximum, which is not scale-free, you have to specify the order yourself. Figure 6.16 shows an example using the *body* dataset again, sorting by the median of the raw data and presenting all variables on the same scale.

If the sorting should only be done on a statistic for a particular subset of the cases, use the appropriate subset in the calculation of the statistic. The following code has been modified to use only the subset for which Age < 30 in specifying the order:

```
m3 <- apply(body1[body1$Ag < 30, 1:24], 2, median, na.rm=TRUE)
```

You will then probably want to emphasise the subset used for sorting using colour (cf. §6.7 below).

These sortings are not scale-free and have been carried out on the original data. This is fine if the original variables have comparable scales, but not always advisable otherwise. To sort on comparable scales, you have to convert the data yourself first as in §6.7 or, if the transformation you want is one of the options available in ggparcoord, use that function to do the work.

For instance, in Figure 6.17 the code orders the pcp axes by the maximum value of the variables standardised by subtracting the mean and dividing by the standard deviation. First of all a pcp of the variables in default order is prepared with that transformation. Then the transformed data are extracted from the plot using a **reshape2** function and ordered. Finally, the pcp with the desired ordering is produced.

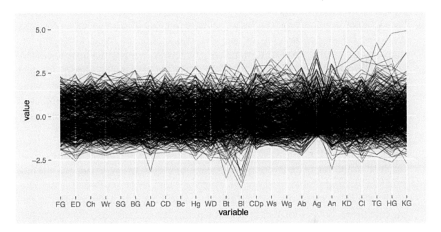

FIGURE 6.17: A pcp of the *body* dataset with the variables sorted by their maximum values after standardisation by mean and standard deviation. Possible outliers are easier to see, particularly the case that is outlying on the two rightmost variables.

```
B1 <- ggparcoord(data = body1, columns=c(1:24), scale="std")
B2 <- acast(B1$data[ ,c(1,3,4)], .ID ~ variable)
m4 <- apply(B2, 2, max, na.rm=TRUE)
m4r <- order(m4)
ggparcoord(data = body1, alphaLines=0.3,
            columns=c(1:24), scale="std", order=m4r)
```

Sorting is a surprisingly powerful tool for data analysis, whether you sort variables or cases or both. Ideally, it should be easier to do for parallel coordinate plots. That would mean presenting a limited range of alternatives in a simple structure and might not cover all the options we would like to have. This is a topic that deserves further thought.

Formatting

Formatting includes another raft of choices and only some are listed here. Mundane decisions like the choice of window size, labelling the axes, or where to put a legend, can all affect what you can see in a display. While this is true for any graphic, it is more important for pcp's, which sometimes need all the window width they can get.

Type of display The standard pcp represents cases by polygonal lines. The individual points can be plotted on each axis with the option `showPoints=TRUE`. This is useful for datasets with few cases; otherwise it only increases the clutter in an already densely packed display. Another option is to plot boxplots for each variable with `boxplot=TRUE`. This can give a feel for the distributions of the variables, especially when there is a large number of cases. Alpha-blending, discussed below, can be used to hide the lines completely or to downplay the lines enough for the boxplots to stand out, while still enabling the lines to convey useful information (cf. Figure 6.9). If the lines are hidden completely, the plot looks like a collection of individual boxplots, just as you might get from drawing boxplots of one variable for different subsets of a dataset. In a pcp boxplot, each case appears in each boxplot, while in a boxplot by subset, each case generally appears only once in the whole plot.

Missings The default in `ggparcoord` is to exclude cases with any missings. That can lead to too many cases being dropped. Other options are to impute missing values. Omitting profile sections for which one end point is missing is not currently possible. As `ggparcoord` decides which cases to exclude after scales are calculated, not before, a case not shown may have determined the scale used.

Aspect ratio Parallel coordinate plots are best drawn wide and moderately high. The appropriate window size depends on the number of variables plotted and the structure of the data. As usual, it is worthwhile experimenting interactively with a few different sizes to see what information is revealed.

Orientation Drawing pcp's horizontally rather than vertically offers the advantage of writing variable names horizontally one above the other, so that no problems arise with overlapping text. An example was shown in Figure 6.3. It is more common to draw pcp's vertically.

Lines Lines can be reformatted in several ways. In principle, their thickness can be varied. In practice, given the large number of lines, they should always be very thin. The most important formatting effects are colour and alpha-blending.

Colour Lines can be coloured in groups by a factor variable or on a continuous shading scale by a numeric variable. Figure 6.18 shows examples of both. In the upper plot, a new binary variable was created to label the 16 Boston areas having the maximum `medv` value of 50 as 'High', and the lines are coloured accordingly. In the lower plot, the lines are coloured according to the case values of `medv`, the median housing value for each area. The cases in the upper plot are ordered to ensure that the group of interest is drawn last.

```
data(Boston, package="MASS")
Boston1 <- within(Boston,
          hmedv <- factor(ifelse(medv == 50,"Top", "Rest")))
Boston1 <- within(Boston1, mlevel <- ifelse(medv==50,1,0.1))
Boston1 <- within(Boston1, medv1 <- medv)
a <- ggparcoord(data = Boston1[order(Boston1$hmedv),],
              columns=c(1:14), groupColumn="hmedv",
              scale="uniminmax", alphaLines="mlevel",
              mapping = aes(size = 1)) + xlab("") + ylab("") +
              theme(axis.ticks.y = element_blank(),
              axis.text.y = element_blank())
b <- ggparcoord(data = Boston1, columns=c(1:14),
              groupColumn="medv1", scale="uniminmax") +
              xlab("") + ylab("") +
              theme(axis.ticks.y = element_blank(),
              axis.text.y = element_blank())
grid.arrange(a,b)
```

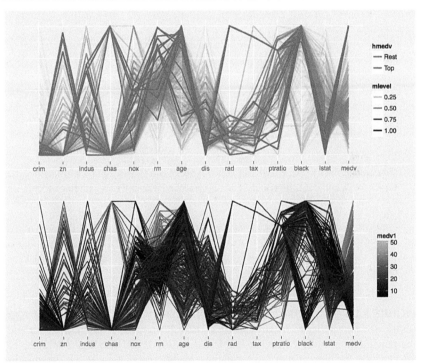

FIGURE 6.18: Plots of the *Boston* dataset with cases coloured by the variable medv, the median home value. In the upper plot, the 16 cases with a value of 50 have been coloured differently and drawn last, so that none of the other 490 cases in the dataset are drawn on top of them. The other cases have been downplayed by using alpha-blending, and all cases have been drawn with a thicker line to help the selected group stand out better. The group differs a little from the rest, but perhaps not as clearly as might have been thought beforehand. In the lower plot, the cases are shaded by their value on the variable medv, an interesting idea technically, although not one that seems to work well for this application.

126   *Graphical Data Analysis with R*

Alpha-blending Parallel coordinate plots suffer from overplotting. If you display
ten variables for 5,000 cases, then you need 50,000 points and 45,000 line segments. Alpha-blending can be used to lessen the problem, although how much
it succeeds in solving it depends on the dataset and on what information is contained in it. The option `alphaLines` in `ggparcoord` can be used, as we have
seen already in Figure 6.2 and Figure 6.10.

To make some groups stand out more than others, `ggparcoord` offers the
possibility of applying a user-specified variable in which the level of alpha-
blending can be set for each case individually. Figure 6.19 shows a pcp of the
Boston dataset in which the 132 cases with the maximum value of `rad` have
been coloured differently and the other cases have been drawn with an alpha-
blending value of 0.1. The order of the variables has also been changed using the
`ggparcoord` option `order`.

```
Boston1 <- Boston1 %>% mutate(
            arad = factor(ifelse(rad < max(rad), 0, 1)),
            aLevel = ifelse(rad < max(rad), 0.1, 1))
ggparcoord(data = Boston1, columns=c(1:14),
            scale="uniminmax", groupColumn= "arad",
            alphaLines="aLevel", order="allClass") +
            xlab("") + ylab("") +
            theme(legend.position = "none",
            axis.ticks.y = element_blank(),
            axis.text.y = element_blank())
```

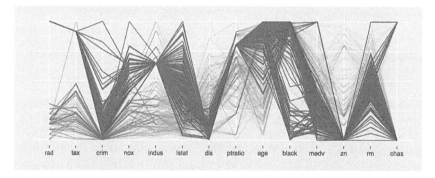

FIGURE 6.19: A pcp of the *Boston* dataset. The cases with the highest value of `rad`
(index of accessibility to radial highways), have been given a different colour and
the remaining cases have been de-emphasised using alpha-blending. The order of
the variables has been changed to accentuate the differences between the two groups
of cases. Interestingly, the five variables in which the `max(rad)` cases all have the
same value, `rad`, `tax`, `ptratio`, `zn`, and `indus`, are not placed together. They
also have lower values for `dis`, higher values for `crim`, tend to have higher values
for `age`, and, with a few exceptions, have lower values for `medv`.

6.8   Modelling and testing for multivariate continuous data

1. Outliers
   As discussed several times in the book, there are no fully satisfactory ways of testing for outliers and you need to consider the individual cases or try some robust approach. If cases are outliers on several variables at once, then that is some evidence for viewing them with suspicion. Interactive graphics are the best approach here, as the order of variables in a pcp determines whether you can recognise such outliers.

2. Regular gaps in the distribution of a variable
   Regular gaps can be due to a measurement feature. For instance, if decathlon performance data are plotted, both the high jump and pole vault events will have gaps, since only certain performances are possible.

3. Clusters of cases
   Occasionally certain groups of cases stand out as having very similar values across a set of variables. This might be tested by comparing that group's means with the rest or by comparing the multivariate distance of the group from the rest.

4. Separated groups
   Sometimes there are clearly separated groups on particular variables. Whether this means anything depends on the values those cases take on other variables, especially categorical ones. Linear models can be useful in assessing the features.

6.9   Parallel coordinate plots and comparing model results

With the computing power that is available nowadays it is becoming more and more common to consider large numbers of models in an analysis. If many different models are fitted to a dataset, it can be a complex matter to compare all the results. Model fitting statistics such as AIC offer a quick filtering mechanism, but are not informative enough for in-depth comparisons. A number of visualisation approaches have been proposed including using a range of parallel coordinate plots to aid the comparisons [Unwin et al., 2003]. Apart from plotting the raw data variables, they suggest plotting the t statistics for comparable parameter estimates with one axis per parameter, and plotting the residuals both by model and by case.

Main points

1. Parallel coordinate plots are a powerful multivariate display, showing all continuous variables at once (e.g., Figures 6.1 and 6.10).

2. There are many pcp display options in addition to the usual formatting options applicable to other graphics: the choice of variables, the order of variables, what scaling is used, and how the axes are aligned (§6.7). Choosing good options is important. Figures 6.14 and 6.19 give examples.

3. pcp's are helpful for exploring and evaluating the results of analyses such as discriminant or cluster analyses (cf. Figure 6.6). They are also useful for checking multivariate outliers, which will be further discussed in §9.3.

4. pcp's are an alternative means for presenting multiple regularly spaced time series and offer the option of a cross-sectional view of the series (§6.5 and Figure 6.9).

5. pcp's are informative for displaying indices and their component parts together (§6.6).

6. You may need to draw several pcp's to uncover the different features in a dataset and other graphics will be useful as well (e.g., the various displays of the *food* dataset in Figures 6.1 to 6.4).

Exercises

1. **Swiss**

   The dataset *swiss* contains a standardized fertility measure and various socioeconomic indicators for each of 47 French-speaking provinces of Switzerland in about 1888. It was already used in Exercise 6 in Chapter 1.

   (a) Draw a parallel coordinate plot of all six variables.
   (b) Are there any cases that might be outliers on one or more variables?
   (c) What can you say about the distribution of the variable `Catholic`?
   (d) Construct a new variable with values 'High' for all provinces with more than 80% Catholics and 'Lower' for the rest. Draw a pcp coloured by the groups of the new variable. How would you describe the provinces with a high level of Catholics?

2. **Pottery**

The package **HSAUR2** includes a dataset on the chemical composition of Romano-British pottery, *pottery*, with 45 cases and 10 variables.

   (a) Draw a pcp of the nine composition variables. What features can you see?
   (b) Make a new variable with the cases with low values on MgO. How are these cases different from the rest on the other variables?
   (c) Colour your original pcp using the site information, kiln. Which kilns can be easily distinguished from the others using which variables?

3. **Olive oils**

The olive oils dataset was introduced in Exercise 8 of Chapter 5.

   (a) Draw a default parallel coordinate plot and describe the various features you can see.
   (b) Draw the same plot and additionally colour the oils by the region they come from. What additional information can you find?
   (c) Discuss which features of the dataset are easier to see with a pcp and which are easier to see with a scatterplot matrix.

4. **Cars**

The dataset *Cars93* was introduced in §5.4. Draw a pcp of the nine variables Price, MPG.city, MPG.highway, Horsepower, RPM, Length, Width, Turn.circle, and Weight.

   (a) What conclusions would you draw from your plot?
   (b) What plot would you draw to compare US cars with non-US cars on these variables? What does the plot tell you about the differences between US cars and the others?
   (c) Is a pcp with unimax scaling informative? Try colouring it by the factor variable Cylinder to gain additional insight.

5. **Bodyfat**

The dataset *bodyfat* is available in the **MMST** package. It provides estimates of the percentage of body fat of 252 men, determined by underwater weighing, and body circumference measurements. The dataset is used as a multiple regression example to see if body fat percentage can be predicted using the other measurements. Draw a parallel coordinate plot for the dataset.

   (a) Are there any outliers? What can you say about them?
   (b) Can you deduce anything about the height variable?
   (c) What can you say about the relationship between the first two variables, *density* and *bodyfat*?
   (d) Do you think the ordering of the variables is sensible? What alternative orderings might be informative?

6. **Exam marks**

   In the package **SMPracticals**, there is a dataset *mathmarks* with the marks out of 100 in five subjects for 88 students. The dataset is fairly old, first appearing in the statistical literature in [Mardia et al., 1979] and it was used in an example at the end of §5.6. It is interesting to note that all students had marks in all subjects. Possibly students who missed an exam were excluded.

   (a) Explore the dataset using pcp's. What information can you uncover and which pcp would you use to present your results to others?

   (b) Apparently the first two exams (`mechanics` and `vectors`) were closed book, while the other three were open book. Draw a pcp with boxplots to see if there is evidence that the students got lower marks on closed-book exams. Is it useful to superimpose the polygonal lines (possibly using alpha-blending) or not?

7. **Wine**

   The *wine* dataset can be found in the packages **gclus**, **MMST**, **pgmm**, and **rattle**. They took the data from the UCI Machine Learning Repository [Bache and Lichman, 2013]. The original source is an Italian software package [Forina et al., 1988]. The version in **pgmm** has about twice as many variables as the others, and the version in **MMST** includes the names of the three classes of wine, rather than the numeric coding that the other versions use.

   (a) Use pcp's to investigate how well the variables separate these classes.

   (b) Are there any outliers?

   (c) Is there evidence of subgroups within the classes?

8. **Boston housing**

   Carry out a cluster analysis of the *Boston* data using Ward's methods (`method=ward.D2`) on standardised variables and choose the four cluster solution. You could present your results in a number of ways:

   (a) with a single pcp of the variables with the case profiles coloured by cluster;

   (b) with several pcp's, one for each set of cluster cases (at the time of writing this will not work with *ggparcoord* if all the cases in a cluster have the same value on a variable when a default or `uniminmax` scaling is used);

   (c) with several pcp's, one for each set of cluster cases and with the remaining cases plotted in the background.

   What are the advantages and disadvantages of the three alternatives? What plot or group of plots would you choose for displaying your clustering results?

9. **Intermission**

   Jackson Pollock's *Convergence* is in the *Albright-Knox Art Gallery* in Buffalo, New York. What can you see in this picture?

7

Studying Multivariate Categorical Data

Most statistical tables are parchingly dry in the reading.

Herman Melville (*Moby Dick*)

Summary

Chapter 7 discusses ways of displaying combinations of categorical variables using various types of mosaicplot.

7.1   Introduction

One-dimensional displays for categorical data like barcharts and piecharts were discussed in Chapter 4. Displays for multivariate categorical data have been discussed in research publications, but have received less attention in applications. The many splendid books on the analysis of categorical data generally include few graphics. The ideal is to display the counts of cases in the multivariate combinations in an informative and easy to interpret way.

When there are several variables, each with a few categories, there can be an unexpectedly large number of possible orderings of variables and categories and it is a challenge to find an effective display. For J nominal variables $\{X_j, j = 1,\ldots,J\}$ with numbers of categories $\{c_j, j = 1,\ldots,J\}$, the variables can be ordered in $J!$ ways and the categories within the variables in $\Pi_{j=1}^{J} c_j!$ ways giving $J!\Pi_{j=1}^{J} c_j!$ orderings in all. In the dataset *Alligator* in the package **vcdExtra** there are $J = 4$ variables: lake (4 categories), sex (2), size (2), and food (5), making $276,480$ orderings in all, an impressively large number considering that there are only 219 observations. As examples will show, there are often compelling arguments for choosing particular orderings, either from the logic of the application or for data-driven reasons.

Now that modern computing power and software have made drawing graphics much easier, you can explore a range of displays to identify the ones that work best. Many types of graphics have been suggested for multivariate categorical data,

131

including mosaicplots, doubledecker plots, fluctuation diagrams, treemaps, association plots, and parallel sets/categorical parallel coordinate plots ([Friendly, 2000], [Meyer et al., 2006], [Kosara et al., 2006], and [Pilhoefer and Unwin, 2013]). In this chapter mosaicplots and their variants will be used. They provide a wide range of possibilities for the display of multivariate categorical data.

7.2 Data on the sinking of the Titanic

```
doubledecker(Survived ~ Sex, data = Titanic,
             gp = gpar(fill = c("grey90", "red")))
doubledecker(Survived ~ Class, data = Titanic,
             gp = gpar(fill = c("grey90", "red")))
```

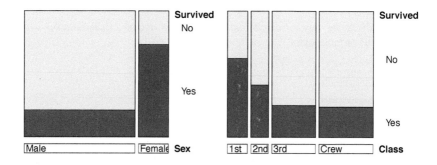

FIGURE 7.1: Survival rates on the Titanic. Females and first class passengers had the best survival chances.

The dataset *Titanic* was used in Chapter 1 Exercise 5 and discussed in §4.3. Barcharts of the dataset's four variables were displayed in Figure 4.4 to give a simple overview. The differing survival rates by groups are of more interest and Figure 7.1 shows spineplots (see §7.3 below) of how they vary by Sex and Class individually. As we all know, a much higher proportion of females survived than males and it is perhaps no surprise that the survival rate was higher for the first class than the second class and that the third class rate was lower again. Does the same class effect apply separately for males and females? A doubledecker plot (§7.4), a generalisation of spineplots, and a form of mosaicplot, can be drawn to investigate.

Figure 7.2 reveals what happened. The four bars to the left show that the male rates do not decline with class and that the males in second class had the lowest survival rate of all groups. The four bars to the right show that for the females, the survival rate declined with class, although of the few females in the crew a high

```
doubledecker(Survived ~ Sex + Class, data = Titanic,
             gp = gpar(fill = c("grey90", "red")))
```

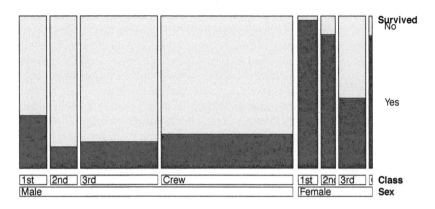

FIGURE 7.2: A doubledecker plot of the Titanic survival rates by class for each sex. The width of each bar is proportional to the number of cases for that sex and class. All female survival rates were higher than all male survival rates and the male survival rate was lowest for second class passengers.

proportion survived. It is fairly easy to see that all female survival rates are higher than all male survival rates. However, if the main aim was to compare female and male rates directly by class, another plot would be better. How to choose which mosaicplot to draw in any given situation is discussed in §7.5.

7.3 What is a mosaicplot?

In a mosaicplot the graphics area is divided up into rectangles proportional in size to the counts of the combinations they represent. Lengths are easier to judge and compare than areas, so it is best to use displays where each rectangle has the same width or the same height.

Mosaicplots are drawn by starting with an empty rectangle to represent the whole dataset, then taking the first variable and dividing the horizontal axis into sections proportional to the sizes of its categories. In the second step each of the rectangles is divided along its vertical axis according to the sizes of the second variable categories, as in the two examples in Figure 7.1. In theory you can continue to divide up the rectangles alternately horizontally and vertically for as many variables as you have. In practice, the display can quickly have too many combinations and it becomes impossible to see what is going on.

Figure 7.3 shows a sequence of default mosaicplots for the *Alligator* dataset mentioned in §7.1, beginning with the `lake` variable and adding the variables `sex`, `size`, and `food` in turn. The category names have been abbreviated to improve the legibility of the labelling, that is always a problem with high-dimensional mosaicplots. The first plot shows that the lakes each provide about the same amount of data; the second that Lake Hancock has more female alligators than the others; the third that generally there are more small alligators than large ones, except for the males in Lakes Oklawaha and Trafford; the fourth plot that the feeding patterns vary. A different ordering of the variables might reveal other information.

The study concerned the eating preferences of alligators in four different lakes, determined by examining the contents of their stomachs. The webpage [Brain, 2004] has this to say about alligators' feeding habits: "Once a week is a typical feeding schedule for alligators living in the wild. Excess calories are stored in fat deposits at the base of the alligator's tail. Incredibly, by burning fat reserves, it is possible for an alligator to last more than two years between feedings." This dataset was first used in one of Agresti's excellent textbooks on categorical data analysis [Agresti, 2007]. The final plot with all four variables displays rectangles for each of the 80 possible combinations and dotted lines for combinations that were not observed. You can see that some combinations did not arise in the dataset, but not a lot else.

This example exemplifies the difficulty of displaying multivariate categorical data effectively and also the difficulty of labelling mosaicplots. In general, there are too many awkwardly aligned combinations to provide a well-defined labelling. The **vcd** package usually does a reasonable job under difficult circumstances.

For a limited number of variables with only a few categories, mosaicplots can work well, although they need careful design. So what is to be done? Given the number of different orderings calculated in §7.1, the various alternative mosaicplots defined in §7.4, and the additional options outlined in §7.6, there are many, many possibilities for finding a better plot and it is necessary to make informed choices and do more than just use a default.

A reasonable strategy is to begin with barcharts of the individual variables to get an initial overview of the dataset. Often some of the features you can discern with difficulty in higher dimensional displays are directly visible in lower dimensional ones. For the *Alligator* dataset you could use code such as

```
data(Alligator, package="vcdExtra")
Alg1a <- aggregate(count~food, data=Alligator, sum)
ggplot(Alg1a, aes(food, count)) + geom_bar(stat="identity")
```

to draw each of the barcharts. If you did, you would see that there were more males than females and more small alligators than large ones, that the numbers were roughly equal for the four lakes, and that fish and invertebrates were clearly the most frequently found foods.

```
data(Alligator, package="vcdExtra")
Alg1 <- Alligator
levels(Alg1$lake) <- abbreviate(levels(Alg1$lake), 3)
levels(Alg1$size) <- abbreviate(levels(Alg1$size), 3)
levels(Alg1$food) <- abbreviate(levels(Alg1$food), 2)
par(mfrow=c(2,2), mar=c(4 ,4, 0.1, 0.1))
mosaicplot(xtabs(count ~ lake, data=Alg1), main="")
mosaicplot(xtabs(count ~ lake + sex, data=Alg1), main="")
mosaicplot(xtabs(count ~ lake + sex + size, data=Alg1),
    main="")
mosaicplot(xtabs(count ~ lake + sex + size + food, data=Alg1),
    main="")
```

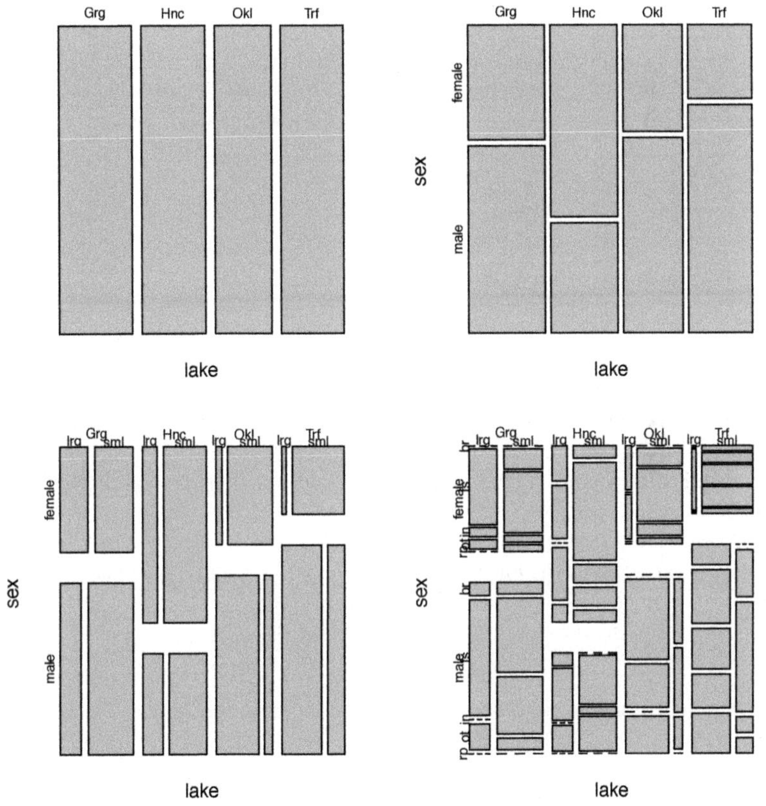

FIGURE 7.3: A sequence of mosaicplots for the *Alligator* dataset. The first plot is a spineplot of the variable `lake`, the second adds the variable `sex`, the third `size`, and the fourth `food`. The category labels have already been abbreviated, so only drawing the plots bigger would alleviate the overplotting of the labels. The first three plots give information on the relative importance of the four lakes in the study, on the sex distribution across the lakes, and on the distribution of alligator sizes by lake and sex. The final plot attempts to present the four-dimensional information.

```
pairs(xtabs(count ~ ., Alligator))
```

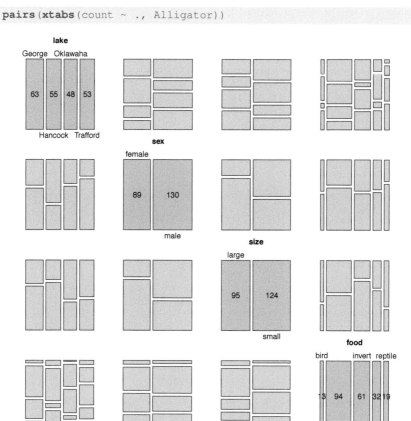

FIGURE 7.4: A plot matrix of barcharts and two-dimensional mosaicplots for the *Alligator* dataset. All pairs of variables barring sex and food appear to be related, since the corresponding bivariate plots do not take the form of a grid.

Alternatively you could use the pairs function in **vcd**, giving Figure 7.4. This produces a matrix display with the barcharts of the individual variables down the diagonal as well as all six mosaicplots both above and below the diagonal. Labelling is kept to a minimum for space reasons and, while efficient, the plot is hard to read. Nevertheless, it can be seen that apart from the two variables sex and food, each pair of variables looks related in one way or another. There are more large males, small alligators eat more invertebrates, the lakes have different patterns of alligators' food preferences. It is always worthwhile to check results in more detail, just to be sure, and so it would be good practice to pick out some of the interesting mosaicplots in Figure 7.4 and draw them individually. Figure 7.5 shows two examples, doubledecker plots of `sex` and `lake` and of `size` and `food`.

At this stage, rather than drawing versions of the four possible three-dimensional mosaicplots it would be practical to incorporate the aim of the study, to see how `food` depends on the other three variables. As `food` has five categories with no particular order a multiple barcharts view (§7.4) could be best. Figure 7.6 shows how *food* depends on `sex` and `size` of the alligators.

```
doubledecker(xtabs(count ~ lake + sex, data = Alligator),
             gp = gpar(fill = c("grey90", "steelblue")))
doubledecker(xtabs(count ~ food + size, data = Alligator),
             gp = gpar(fill = c("grey90", "tomato")))
```

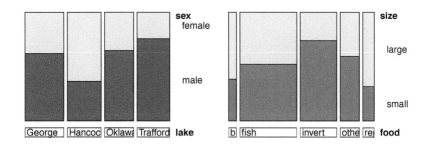

FIGURE 7.5: Doubledecker plots for the *Alligator* dataset. From the left-hand plot you can see that the `sex` distribution varies by `lake` and from the right-hand plot that the variables `size` and `food` are associated.

```
Alg2 <- aggregate(count~sex + food + size, data=Alligator, sum)
ggplot(Alg2, aes(food, count, fill=sex)) +
      geom_bar(stat = "identity") +
      facet_grid(size ~ sex) + theme(legend.position="none")
```

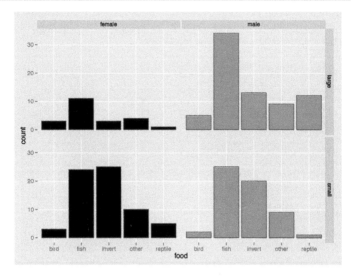

FIGURE 7.6: A multiple barcharts view showing how `food` depends on `sex` and
`size` in the *Alligator* dataset. Males and females of the same size have similar patterns, but large alligators prefer fish while smaller ones like invertebrates too.

A further step would be to include the `lake` variable and Figure 7.7 shows a
multiple barcharts view with the `food` distributions for the four combinations of
`sex` and `size` for each of the four lakes separately.

Figure 7.7 is obviously a better choice of display than the default mosaicplot for
all four variables in Figure 7.3, as it is much easier to recognise features of the dataset.
While the differences between the lakes are substantial, it is necessary to be careful in
drawing too firm conclusions given the small numbers involved. Loglinear modelling
would be helpful in determining which features of the data have strong support in
the data, while graphical displays provide an appreciation of the raw data and an
understanding of model results. As always, graphics and modelling complement each
other.

To improve the display clarity of mosaicplots, narrow gaps are left between the
category rectangles. With two variables you have a kind of barchart called a spineplot
[Hummel, 1996]; with spineplots there are no gaps between the categories of the
second variable. Without gaps the conditional rates of the second variable, $P(B|A)$,
are directly displayed. As in a barchart, the areas of the bars are proportional to the
counts, but instead of the bars having equal widths and differing heights, they all have
equal heights and differing widths. Spineplots were designed primarily for interactive
graphics to display rates by category for a second linked variable. Gaps are useful if

```
ggplot(Alg1, aes(food, count, fill=sex)) +
        geom_bar(stat = "identity") +
        facet_grid(lake ~ sex + size) +
        theme(legend.position="none")
```

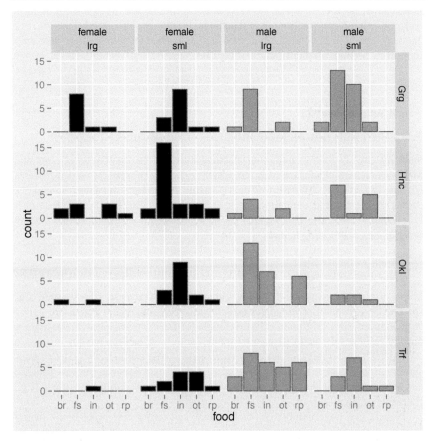

FIGURE 7.7: A multiple barcharts view showing how `food` depends on `sex`, `size` and `lake` in the *Alligator* dataset. There are different patterns for the four lakes, but numbers are quite small.

you are to divide the resulting rectangles according to the category proportions of a third variable.

The construction of mosaicplots is hierarchical and the order of the variables has a big impact on the display. There are other options to consider as well. Instead of the classical alternating horizontal and vertical divisions, mainly horizontal divisions might be used, as is the case with doubledecker plots (cf. Figure 7.2), or indeed any sequence of horizontal and vertical divisions might be used.

Classical mosaicplots go back to the paper [Hartigan and Kleiner, 1981] and aim to make the most efficient use of the space available. This means that the area representing each case is maximised, while the ease of interpretation suffers, as the rectangles representing the combinations are not aligned for direct comparison. Even with a small number of combinations a classical mosaicplot may prove to be ineffective. With a large number of combinations, no mosaicplot is likely to work. For anything from 4 to 24 combinations, mosaicplots can be very good. Above that, their usefulness depends on the structure of the data.

7.4   Different mosaicplots for different questions of interest

We have already seen examples of classical mosaicplots, doubledecker plots and multiple barcharts. There are other alternatives too. Each mosaicplot variant presents the data in a slightly different way and is useful for different purposes. Sometimes counts are emphasised, sometimes rates, and sometimes distributions. In all cases it is essential to remember that we are looking at conditional distributions and that the order of the conditioning matters.

Which subgroups appear most often? (Fluctuation diagrams, Figure 7.8)

In this display, the axes are divided up equally for each category of a variable irrespective of size, and the rectangles representing the combination counts are drawn at their appropriate grid points. The rectangle for the combination with the highest count determines the scale, given the variables and the size of the window.

Fluctuation diagrams are good for representing large contingency tables or transition matrices, where there is no reason to differentiate between the row variable and the column variable. Another application is displaying confusion matrices (e.g., [Pilhoefer et al., 2012]).

If there is a large number of combinations and only a few occur at all, then a fluctuation diagram is valuable for revealing this information and for identifying categorical clusters. Figure 7.8 shows a fluctuation diagram for eight binary variables of the *Zoo* dataset in the package **seriation**. Several distinct groupings stand out and there are many feature combinations for which there are no animals in the dataset.

Fluctuation diagrams can be drawn in R using the `fluctile` function in the package **extracat**. You could compare this fluctuation diagram with a mosaicplot of the same data using the code

```
data(Zoo, package="seriation")
mosaic(table(select(Zoo, c(hair, eggs:backbone))))
```

```
data(Zoo, package="seriation")
fluctile(table(select(Zoo, c(hair, eggs:backbone))),
         label=FALSE)
```

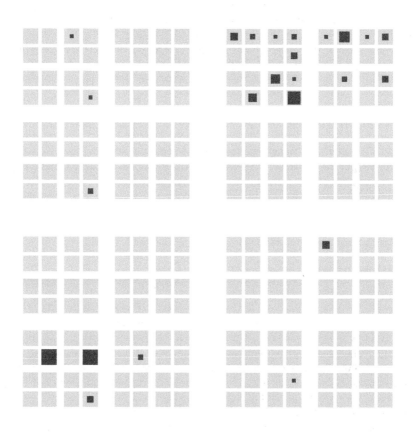

FIGURE 7.8: A fluctuation diagram of eight binary variables (hair, eggs, milk, airborne, aquatic, predator, toothed, backbone) for the 101 cases of the *Zoo* dataset in the package **seriation**. A few combinations arise relatively often, most not at all.

Comparing rates across subgroups
(Doubledecker plots, Figure 7.9)

If only horizontal divisions are used for all but the last variable, you get a doubledecker plot, which is a multivariate spineplot. Each bar has the same height and therefore both its area and its width are proportional to its count. The plot's name comes from how it sometimes looks when you colour the bars by a final binary dependent variable. This is shown in Figure 7.2 for the Titanic dataset. Doubledecker plots are excellent for comparing rates across all groups (the heights in the shaded sections of the bars), while conveying some information on the relative sizes of the groups.

Another example, this time with an ordinal dependent variable, can be seen in Figure 7.9 for the *Arthritis* dataset in the package **vcd**. Improvement can be "marked", "some", or "none" and is dependent on Treatment and Sex. Note how the ordering of the variables, the ordering of the Treatment categories, and the direction of splitting have all been chosen to reflect a clear pattern in the data. Treated patients did better than the placebo patients and females did better than males within those groups. Hardly any males in the placebo group showed any improvement at all.

Doubledecker plots can be drawn in R using the `doubledecker` function in **vcd** or the `rmb` function in **extracat**.

```
data(Arthritis)
Arthritis <- within(Arthritis,
                    Treatment1 <- factor(Treatment,
                    levels=levels(Treatment)[c(2,1)]))
mosaic(Improved ~ Treatment1 + Sex, data = Arthritis,
       direction = c("v", "v", "h"), zero_size = 0)
```

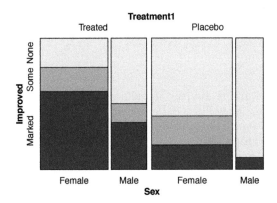

FIGURE 7.9: A doubledecker plot of Improvement for arthritis patients conditional on `Treatment` and `Sex`. The treated group did better than the placebo group and females did better than males.

How do patterns compare by subgroup?
(Multiple barcharts, Figure 7.10)

Complex contingency tables with several rows and columns can be awkward to work with. Tests based on χ^2 statistics may be highly significant, but how should the results be presented? Fluctuation diagrams, treating the row and column variables equally can be effective, although comparisons of rectangular areas are harder than comparisons of lengths.

If one variable can be regarded as dependent and the other explanatory, then multiple barcharts are good. They show the distributions of the dependent variable conditional on the explanatory one. Consider the mosaicplots for the two variables `lake` and `food` for the *Alligator* dataset in the bottom left and top right corners of Figure 7.4 and compare them with the multiple barcharts version in Figure 7.10. Note that the barcharts have been aligned vertically to make comparing distributions easier. Aligning the barcharts horizontally taking up less space would not be as effective.

```
Allig3 <- aggregate(count~food + lake, data=Alligator, sum)
ggplot(Allig3, aes(food, count)) +
       geom_bar(stat = "identity") + facet_grid(lake ~.)
```

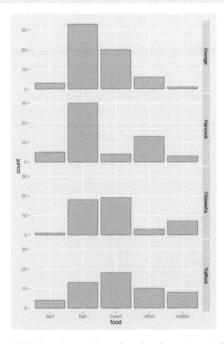

FIGURE 7.10: A multiple barcharts view showing how `food` depends on `lake` in the *Alligator* dataset. The distributions for Lake George and Lake Hancock are quite different from those for the other two lakes.

If there is a final dependent variable that has several categories, then colour can be used to fill the rectangles representing the explanatory variable combinations, and what you get is a stacked barchart display. For certain data patterns this will work well, cf. Figure 7.9. More commonly, the result is difficult to interpret, as there is no cumulative pattern, and so one alternative is to use a grid structure for the explanatory variables and to plot the final dependent variable as a barchart. The resulting display is a trellis plot as in Figures 7.6 and 7.7. These plots can be drawn using the `rmb` function in **extracat** or with **ggplot2** or **lattice**.

Just showing the rates
(Same binsize plots, lower plot in Figure 7.11)

This form is different from the others in that each combination is represented by a rectangle of the same size. Combinations with zero counts are drawn empty. This plot is useful for two reasons, firstly for showing the pattern of combinations which do not arise in the data, the missings, and secondly for comparing rates of a binary dependent variable. Rates can be seen by colouring the rectangles up to a height equivalent to the appropriate values. The disadvantage of the plot is the lack of count information, but when the conditioning groups are of equal size or all large this is irrelevant.

The upper plot of Figure 7.11 is a mosaicplot of the rates of high satisfaction for the 24 combinations of the three variables `Infl`, `Type`, and `Cont` in the *housing* dataset. The order of variables and their alignment (whether horizontal or vertical) has been chosen to emphasise the pattern of generally increasing rates of high satisfaction for increasing influence given contact and type. Because the groups have different sizes, they are not aligned, which makes it more difficult to see the pattern. The lower plot shows the identical data in a same binsize plot. Now the rate patterns are easier to compare, while the differences in group sizes are ignored. Same binsize plots with highlighting can be drawn in R using the `rmb` function in the package **extracat**, using the `freq.trans = "const"` option. The coding and options for the two plots are rather different, as they are drawn with different packages, **vcd** and **extracat**. The various ways R packages go about achieving their goals can be frustrating as well as fruitful.

```
data(housing, package="MASS")
mosaic(xtabs(Freq ~ Cont + Type + Infl + Sat, data = housing),
       direction = c("h", "v", "v", "h"),
       gp = gpar(fill = c("grey", "grey","red")),
       spacing = spacing_highlighting)
```

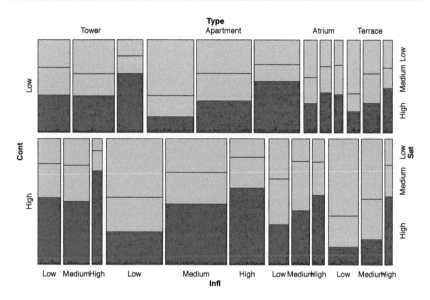

```
rmb(formula = ~Type+Cont+Infl+Sat, data = housing, cat.ord = 3,
    spine = TRUE, freq.trans = "const")
```

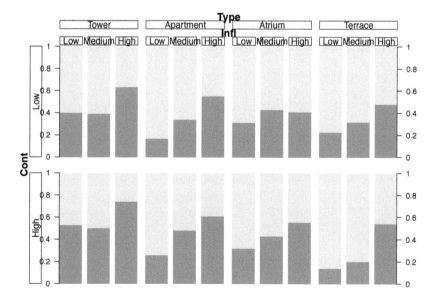

FIGURE 7.11: Rates of high satisfaction for the housing dataset in the **MASS** package. The upper plot is a mosaicplot, which shows the different group sizes, and the lower one a same binsize plot, which only shows the rates.

Comparing patterns for subgroups of very different frequencies (Relative multiple barcharts, *rmb*, Figure 7.12)

Like trellis plots, all subplots in a multiple barchart have the same vertical scale. If the underlying groups are of very different sizes, some of the individual plots will have too small a height to see any pattern. The *rmb plots* in the package **extracat** offer two solutions.

One solution is to adjust the horizontal scale of each subplot, so that the total area of the bars in the plot is proportional to the sample size and the full range of the vertical scale is used. This results in barcharts from small groups being very thin. The distributions represented by the barcharts formerly had comparable horizontal scales but not comparable vertical scales, with this approach it is vice versa.

The other alternative offered is to draw all the barcharts with individual vertical scales and use the colour intensity of the plots to represent the group sizes. Plots from small groups are drawn with lighter shadings than plots from large groups and this can be very helpful when it is the distributions that you want to compare and the absolute values are not so relevant.

The two options are shown for the *housing* dataset in Figure 7.12. There is evidence that satisfaction increases with increasing influence, that within the influence subgroups satisfaction is slightly higher with more contact, and that the tower group have higher satisfaction levels than the other three housing types.

```
data(housing, package="MASS")
rmb(formula = ~Type+Infl+Cont+Sat, data = housing,
    col.vars = c(FALSE,TRUE,TRUE,FALSE),
    label.opt = list(abbrev = 3, yaxis=FALSE))
```

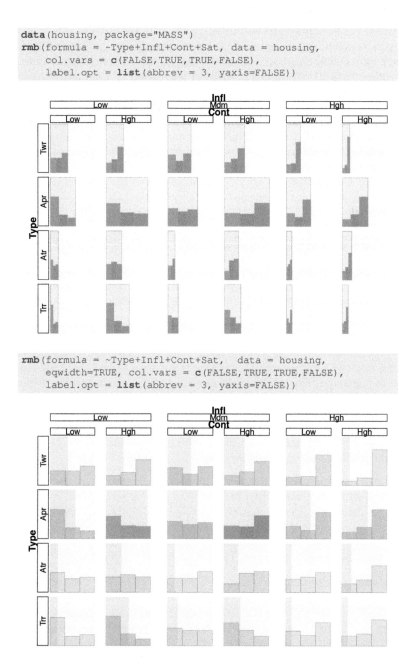

```
rmb(formula = ~Type+Infl+Cont+Sat,  data = housing,
    eqwidth=TRUE, col.vars = c(FALSE,TRUE,TRUE,FALSE),
    label.opt = list(abbrev = 3, yaxis=FALSE))
```

FIGURE 7.12: *rmb plots* of the *housing* dataset. In the upper plot each barchart has the same vertical scale, but its own horizontal scale, and its total area reflects the size of the group in that cell. In the lower plot each barchart has the same vertical and horizontal scales and the intensity of colouring reflects the cell group size.

Graphics supporting modelling (Residual plots)

Mosaicplots can be very helpful for displaying raw data and they can also be used
to support modelling. The rectangles may be drawn with sizes proportional to the
expected values under a particular model and then coloured or part-filled according
to the model residuals. The resulting display is a good way to see if the model fits
and, if it doesn't, where the problems lie. Some recommend drawing the rectangles
with sizes proportional to the raw data values, so that the shape of the plot remains
unchanged, whichever model is fitted. However, this can mean that there are rectangles with zero size, making it difficult to plot residuals. Using expected values is
more consistent with other statistical plots. The **vcd** package includes examples of
applying mosaicplots in modelling and offers some alternative displays.

7.5   Which mosaicplot is the right one?

There are some obvious principles to follow in choosing a suitable mosaicplot form
and structure, but that does not mean that you can find an 'optimal' plot. It is best
to look at several different ones and consider using more than one to display all the
information in a dataset.

Choice of plot form *Classic mosaicplots* are good for ordinal dependent variables
as you can see cumulative patterns if they exist. For dependent variables with
more than two categories *multiple barcharts* are often a good choice. If there
are very many possible combinations, then a *fluctuation diagram* will be helpful
for identifying clusters in the data. *Doubledecker plots* are good for comparing
rates for a binary dependent variable across all possible groupings. *Same binsize
plots* are useful for identifying missing combinations and for comparing rates
when all groups are much the same size, while *rmb plots* are best for comparing
conditional distributions for groups of very different sizes.

Choice of ordering of variables A binary dependent variable should be the last
variable in the plot and is best included as colouring in a vertical split, just
like highlighting in interactive graphics, rather than as a separate variable. Variable ordering establishes what conditional relationships are shown. Figure 7.2
shows survival rates for Class within Sex for the *Titanic* dataset. Another ordering would be better for comparing survival rates by Sex within Class:

```
doubledecker(Survived ~ Class + Sex, data = Titanic)
```

Sometimes there is a natural hierarchical ordering that should be used, otherwise
it can be sensible to use data-driven orderings, if they lead to recognisable patterns. An example can be seen in Figure 7.11 for the *housing* dataset, where the
ordering Cont, Type, Infl was chosen because of the rising pattern that is
revealed.

Choice of ordering of variable categories The categories of ordinal variables must be kept in the correct sequence, either increasing or decreasing. If there is no sensible default ordering for the categories of a nominal variable, then ordering by frequency is often best.

Choice of display form Mosaicplots are complex and need a lot of space, too large is better than too small. The aspect ratio chosen can be crucial in determining what can be seen in a plot and various sizes should be tried. Restrained use of colour can be valuable for emphasising particular features. Labelling is difficult, and so captions and additional annotations can be more important for mosaicplots than for other kinds of plot.

7.6 Additional options

Aspect ratio All graphics displays look different when the window aspect ratio is varied and mosaicplots are especially sensitive. For fluctuation diagrams and the same binsize variant, varying the aspect ratio to produce squarish rectangles seems to work best. This could be said to be related to Cleveland's principle for line charts [Cleveland, 1993], that the slopes should be around 45°, as the rectangle diagonals are then at 45°. For the other forms, taller, thinner rectangles are better. This is in line with another of Cleveland's recommendations, given we want to compare lengths. Deciding what aspect ratio is most informative can be a subjective judgement and so it is a sensible plan to experiment a little, resizing the window with the mouse, to find the most effective view.

Gaps between rectangles The most efficient use of space would be to leave no gaps between the rectangles (as treemaps [Johnson and Shneiderman, 1991] do). This makes comparisons and alignment difficult, so it is usual to leave gaps between the categories whenever a split is made, and it is possible to vary the gap size according to the order of the variables. The higher in the hierarchy the variable appears, the larger the gaps. The choice of gap sizes and how much they are reduced as you go down the hierarchy influence the way the graphic looks and how easy it is to recognise features. Unusually, this is not a situation in which it is easy to experiment. More research is needed.

Rotation Rotating individual variables (i.e. splitting vertically instead of horizontally and vice versa) alters the display layout and can aid interpretation. If there are three binary variables and one variable with eight categories, it will usually be better to split the three binary variables in the same direction. Sometimes it can be helpful to rotate the whole plot at once. The same effect can be achieved by rotating each variable individually. As far as possible layouts should be chosen to emphasise the comparisons you want to make.

Censored zooming There are two kinds of censored zooming to aid the understanding of fluctuation diagrams and multiple barcharts. Censored zooming works best as an interactive tool.

    Ceiling-censored When there are many possible combinations, some will have zero counts, some will have very low counts, and it becomes difficult to distinguish between them. This is immaterial if interest is only in the large groupings, but often categorical outliers can convey important information or it can be useful to be able to examine differences between small groups. With ceiling-censored zooming, a maximum rectangle size is set and any combinations with larger counts are drawn to that size. This magnifies the smaller combinations, while limiting the larger ones. For example, we can convert the Titanic dataset to a data frame with count variable `Freq` and then construct a new count variable with a specified ceiling in the following way:

```
t1 <- as.data.frame(Titanic)
ceil <- 100
t1$Fc <- ifelse(t1$Freq > ceil, ceil, t1$Freq)
```

    and then the plot can be drawn with the new count variable.

    Floor-censored When interest is concentrated on larger groups, combinations with small frequencies may distract rather than help. They can be suppressed by only displaying cells with counts larger than a specified floor. The preparatory code would be

```
floor <- 5
t1$Ff <- ifelse(t1$Freq < floor, 0, t1$Freq)
```

Colour Colour is good for displaying rates in subgroups. It is also valuable for displaying residuals, using separate colours for positive and negative residuals, and for emphasising particular subgroups. All plots can be improved with a judicious use of colour while many plots can be made worse by an injudicious use of colour. Colour should emphasise information or add to the attractiveness of a graphic, there has to be a definite reason for using it.

7.7 Modelling and testing for multivariate categorical data

1. Contingency tables
   The standard for checking the association of two categorical variables is the χ^2–test. In some situations Fisher's exact test is appropriate. There are also tests for special situations like McNemar's and statistics for certain kinds of sets of tables like Mantel-Haenszel.

2. Associations between categorical variables
   If there is a small number of variables with only a few categories and with not too many sparsely occupied combinations, then loglinear models can be used to assess independence structures amongst the variables. Loglinear models can be difficult to interpret and there are usually several that might be considered acceptable. If the necessary conditions hold, and loglinear models can be fitted, then mosaicplots are helpful in understanding the results, both in displays of the raw data and in displays of model residuals.

3. Binary dependent variables
   For a binary dependent variable, logistic regression is a good approach. For an ordinal dependent variable with more than two categories, cumulative link models can be used, as for instance implemented in the R package **ordinal**.

Main points

1. It is difficult to display multivariate categorical data (§7.3).

2. Start with low-dimensional plots to get to know the basic data structure in a dataset (Figures 7.3 and 7.4).

3. Mosaicplots and their variants offer many good alternatives (§7.4).

4. The order of variables in a plot and the orderings of the categories of variables strongly influence what information can be found (§7.5).

5. There are many different kinds of mosaicplots and many different options for them (§7.6). Try out several plots and be prepared to use more than one.

Exercises

1. **Cancer**

   The **vcdExtra** package includes a data table for an old breast cancer study on the survival or death of 474 patients, *Cancer*.

   (a) Convert the data to a data frame and draw plots to compare the survival rates firstly by degree of malignancy and secondly by diagnostic center.

   (b) Which plot would you draw to compare survival rates by both degree of malignancy and diagnostic center? Does the order of explanatory variables matter?

2. **Titanic**

   None of the chapter's graphics for the *Titanic* dataset make use of the Age variable. What mosaicplot would you draw to include this variable and what are your conclusions?

3. **Death penalty**

   A summary table of death penalty verdicts in Florida over a number of years is reported in [Agresti, 2007]. There are three binary variables, whether the victim was black or white, whether the defendant was black or white, and whether the verdict was the death penalty. The dataset can be found in the packages **asbio** (*death.penalty*) and **catdata** (*deathpenalty*). What conclusions would you draw from the data? Choose an appropriate mosaicplot to support your argument.

4. **Berkeley admissions**

   The dataset on graduate admissions to Berkeley was introduced in §1.3 with barcharts of the three variables, Dept (Department), Gender, and Admit. Draw plots to show how Admit varies by Dept and Gender separately and then a plot showing how it varies by the different combinations of Dept and Gender.

5. **Airline arrivals**

   The dataset *airlineArrival* in the package **fastR** includes details of 11000 arrivals at five different airports for two airlines. Compare a spineplot of the rate of delay by airline with a mosaicplot of the rates of delay by airline for each airport separately. Which mosaicplot is most effective here? (This is an example of Simpson's paradox. The data were originally reported in [Barnett, 1994].)

6. **Knowledge of cancer**

   The **vcdExtra** package includes the dataset *Dyke* about how 1729 survey respondents' knowledge of cancer depended on whether they listened to the radio, read newspapers, did 'solid' reading, or attended lectures.

   (a) What plots would you draw to present the information in this dataset of five binary variables?

   (b) If you were restricted to one plot showing all the variables, which one would you choose and why?

7. **Punishment**

   A dataset on attitudes to corporal punishment of children in Denmark, *Punishment*, is provided in the **vcd** package. What plot or plots would you recommend for analysing this dataset and how would you summarise the information from the dataset?

8. **Hair and eye colour**

   The *HairEyeColor* dataset in R has often been used to illustrate mosaicplots. Data on hair and eye colour of 592 students were published by Snee [Snee, 1974].

   (a) Compare a classical mosaicplot with a fluctuation diagram and with a multiple barcharts version for studying the association between hair colour and eye colour. What information do you find and which plot displays it best?
   (b) How important are the orderings of the categories for the two variables?
   (c) The data were later extended by Friendly "for didactic purposes" (R Help), where, as [Friendly, 1994] puts it, "the division by sex is contrived". Which plot would you choose for looking at hair and eye colour by sex and does the variable `sex` make a difference?

9. **Clothing and intelligence**

   A classic example of a larger contingency table is the *Gilby* dataset in the package **vcdExtra**. In his 1911 paper Gilby [Gilby, 1911] reported on a study of 1725 children's standard of clothing (5 categories with the worst two combined from worst/insufficient to very well clad) and teachers' rating of their intelligence (7 categories combined into 6 from mentally defective/slow dull to very able). The χ^2 statistic for the table is extremely significant and we can guess why, but you cannot easily see from the table. Compare a classic mosaicplot, a spineplot, a fluctuation diagram, and a multiple barcharts view for these data. Which do you think conveys the information best?

10. **Dinosaurs in North Dakota and Montana**

    The *HCD* dataset in the **MCPAN** package contains counts of dinosaur families in three stratigraphic levels of the Cretaceous period in the Hell Creek formation in North Dakota and Montana. The table in R is based on a subset of the data used in [Sheehan et al., 1991], reported in [Rogers and Hsu, 2001].

    (a) Draw barcharts of the numbers found at each of the three levels and of the counts for each family. (You may find it helpful to reformat the dataset using the package **reshape2** or in some other way.)
    (b) Choose a suitable mosaicplot to display the data for families by level. The aim of the original study [Sheehan et al., 1991] was to study whether dinosaur family diversity changed across levels.

11. **Intermission**

    Piet Mondrian's *Broadway Boogie Woogie* is in the *Museum of Modern Art* in New York. Does the title complement the picture?

8

Getting an Overview

Diagrams prove nothing, but bring outstanding features readily to the eye.

R. A. Fisher ([Fisher, 1925])

Summary

Chapter 8 discusses getting an initial overall view of a dataset.

8.1   Introduction

When you first start work on a dataset it is important to learn what variables it includes and what the data are like. There will usually be some initial analysis goals, but it is still necessary to look over the dataset to ensure that you know what you are dealing with. There could be issues of data quality, perhaps missing values and outliers (discussed in the next chapter), and there could just be some surprising basic statistics.

There are several different functions in R for showing what is in a dataset. You can show the whole dataset (inadvisable for even moderately sized datasets), display only the first few lines (using the function `head`), or just list the variables, each with their type and a few values (using the function `str`). You can summarise the dataset using the function `summary` (**base** R) or the function `describe` (**Hmisc**) or the function `whatis` (**YaleToolkit**). There are doubtless other statistical alternatives. Plotting the dataset is the alternative, complementary approach for getting an overview and the one we will be concentrating on in this chapter.

As an example, consider the dataset *HI* from the package **Ecdat** on the effect of health insurance on women's working hours. The data were analysed in [Olsen, 1998]. First information can be gained with `str` (the results are shown on the next page) and some simple graphics (Figures 8.1 and 8.2).

```
data(HI, package="Ecdat")
str(HI)
```

```
#    'data.frame': 22272 obs. of  13 variables:
#    $ whrswk : int 0 50 40 40 0 ...
#    $ hhi : Factor w/ 2 levels "no","yes": 1 1 2 1 2 ...
#    $ whi : Factor w/ 2 levels "no","yes": 1 2 1 2 1 ...
#    $ hhi2 : Factor w/ 2 levels "no","yes": 1 1 2 2 2 ...
#    $ education : Ord.factor w/ 6 levels
#       "<9years"<"9-11years"<..: 4 4 3 4 2 ...
#    $ race : Factor w/ 3 levels "white","black",..: 1 1 1 1 1
#    ...
#    $ hispanic : Factor w/ 2 levels "no","yes": 1 1 1 1 1 ...
#    $ experience: num 13 24 43 17 44.5 ...
#    $ kidslt6 : int 2 0 0 0 0 ...
#    $ kids618 : int 1 1 0 1 0 ...
#    $ husby : num 12 1.2 ...
#    $ region : Factor w/ 4 levels "other","northcentral",..: 2
#       2 2 2 ...
#    $ wght : int 214986 210119 219955 210317 219955 ...
```

There are seven factors (one of which, education, is ordered), two numeric variables and four integer ones. The two kids variables must be discrete and limited, as drawing up tables or plotting them would confirm.

Figures 8.1 and 8.2 show histograms of the four continuous variables and barcharts of the remaining variables respectively. The histograms' display is drawn by forming a new dataset, stacking the continuous variables, so that all the plots can be drawn together with one line of code. (This could be done for the barcharts too, but then any category ordering information for nominal variables would be lost.)

The hours worked by the women have two distinct modes, which on closer inspection of the data turn out to be 0 hours (i.e., not working) and 40 hours. The odd shape of the histogram of experience is due to the default binwidth, but the overall shape is clear. The extended axis below zero is surprising and growing the window vertically would show that there are a very few cases with a value of -1. (These arise because the variable is defined as years of potential work experience = age − years of education − 5.) Both the last two variables, husband's income and sampling weight, are skewed to the right. Sampling weight distributions are often skew with a few cases having exceptionally large weights. The mode in the salary histogram at 100 turns out to be real, with 478 husbands' incomes reported as $99,999 a year! This can be checked with something like

```
with(HI[HI$husby > 98 & HI$husby < 102 ,], table(husby))
```

```
library(reshape2)
HIvs <- c("whrswk", "experience", "husby", "wght")
HIs <- melt(HI[, HIvs], value.name = "HIx",
          variable.name = "HIvars")
ggplot(HIs, aes(HIx)) + geom_histogram() +
      facet_wrap(~ HIvars, scales = "free") +
      xlab("") + ylab("")
```

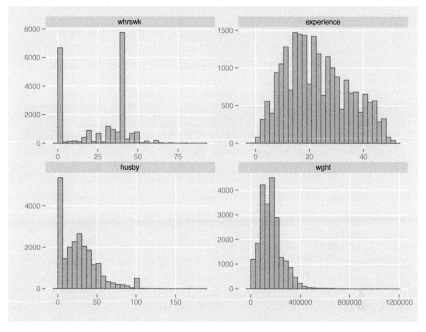

FIGURE 8.1: Histograms of the four continuous variables in the *HI* dataset from **Ecdat**. Women mostly work 40 hours per week or do not work at all. Potential years of experience range from 0 to around 50. A large group of husbands have little or no reported income. The distribution of case weights is skewed to the right.

The barcharts in Figure 8.2 show the insurance variables, information on education and race, the distributions of young and older children, and the distribution by region. The variables are selected by finding out which ones have a limited number of unique values. The weighting variable has been converted to a percentage to make the scales readable.

```
uniqv <- function(x) length(unique(x)) < 20
vcs <- names(HI)[sapply(HI, uniqv)]
par(mfrow = n2mfrow(length(vcs)))
relativeWeight <- with(HI, wght/sum(as.numeric(wght))*100)
for(v in vcs)
    barplot(tapply(relativeWeight, HI[[v]], sum), main = v)
```

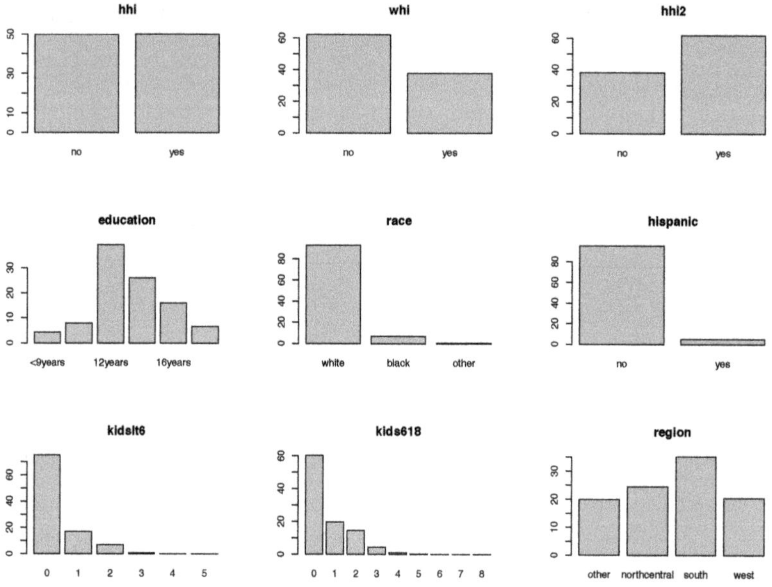

FIGURE 8.2: Barcharts of the nine other variables in the *HI* dataset in **Ecdat**. The top row shows that about half the women are covered by their husband's insurance and a minority by their own, and that an equivalent minority of husbands have no job health insurance. The middle row shows that most women have twelve or more years of education and that there are few blacks and Hispanics in the study. The final plot at the bottom right shows that more participants were from the South.

Sometimes it is obvious which variables should be treated as continuous and which as categorical or discrete, sometimes it is ambiguous. If an age variable takes values from 0 to 100, then it makes sense to treat it as continuous, while if there are only integer values between 20 and 30 it might be better to treat it as discrete. If you guess wrongly, then draw another plot. The whole point of initial overviews is to get a feel for the data, not to draw perfect pictures. They can come later when you have learned what information in the data is worth presenting.

8.2   Many individual displays

Instead of jumping in and producing lists of summaries (as with summary) or a large matrix of primarily scatterplots (as with plot), another approach is to begin with the basics and work up step by step. Knowing what variables of what kinds there are and how many cases is a pretty good start and can be achieved using str, as was done in §8.1. The GDA approach suggested here is to split the variables into two groups, plotting categorical and discrete variables as barcharts, while plotting the other variables, where possible, as histograms. Any special variable types left over (e.g., dates) should be dealt with separately.

Plots of individual variables give a quick view of the variable distributions and of any features that stand out. Plotting variables in groups, all histograms together and all barcharts together, is quicker than plotting them one by one and organises them neatly. The Boston housing dataset was already examined in Figure 3.9, treating all variables as continuous. In fact the dataset's 14 variables include one binary, one discrete, and twelve numeric. In Figure 8.3 the binary and discrete variables have been plotted and in Figure 8.4 all the continuous variables.

```
data(Boston, package="MASS")
par(mfrow=c(1,2))
for (i in c("chas", "rad")) {
    barplot(table(Boston[, i]),
    main=(paste("Barchart of", i)))
}
```

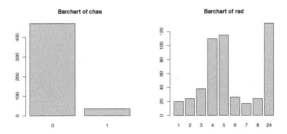

FIGURE 8.3: Barcharts of the variables chas (whether the tract bounds the Charles River or not) and rad (index of accessibility to radial highways) from the *Boston* dataset.

Although neither of these is ideal for the individual variables (for instance the histogram of medv misses the collection of areas with medv= 50, which was identified in §3.3), they do offer some direct insights (variable definitions can be found on the dataset's R help page):

1. There are few areas with chas = 1.

2. Three values, 4, 5, and 24, dominate the variable rad. The value 24 is quite separate from the others in value (the plot fails to show that clearly because barcharts treat a discrete variable's numbers as category names).

3. The variables tax and indus have gaps that could be investigated.

4. Three variables are highly skewed (crim, zn, and black).

5. And there are suggestive features in other variables to be studied more closely.

```
vs1 <- !(names(Boston) %in% c("chas","rad"))
grs <- n2mfrow(sum(as.numeric(vs1)))
par(mfrow=grs)
for (i in names(Boston)[vs1]) {
    hist(Boston[,i], col="grey70", xlab="", ylab="",
    main=(paste("Histogram of", i)))
    }
```

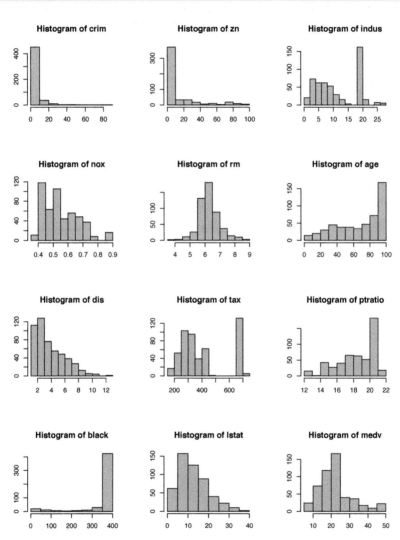

FIGURE 8.4: Histograms of the twelve continuous variables from the *Boston* dataset.

8.3   Multivariate overviews

Scatterplot matrices have already been mentioned as a way of studying relationships between variables and they can be very effective. There are other possibilities as well. Apart from parallel coordinate plots and trellis plots there are tablelens plots, heatmaps, and glyphs. The following subsections discuss examples of several of these alternatives for the *Boston* dataset (variable definitions can be found on the dataset's R help page).

Scatterplot matrices

The default scatterplot matrix (well, default except for using points rather than open circles) shown in Figure 8.5 is surprisingly informative, even with 14 variables. A number of features stand out:

1. The curious either/or relationship between crim and zn.

2. The dependencies of medv on lstat and rm.

3. The fact that tax, ptratio, zn, and indus have only single values for the extreme value 24 of rad.

4. The associations between age and dis and nox.

5. And some potential bivariate outliers such as the point in lstat and age or the set of high values of nox, which have the same value for rad, tax, and ptratio.

As a next step you could consider adding univariate displays of the variables down the diagonal or colouring cases by membership of some subgroup. Figure 8.5 is just intended to provide a first quick look to help you to decide how to proceed. It may be a complex graphic, but it has a straightforward structure and is easy to draw. We should take advantage of the power that software can offer us nowadays.

```
plot(Boston, pch=16)
```

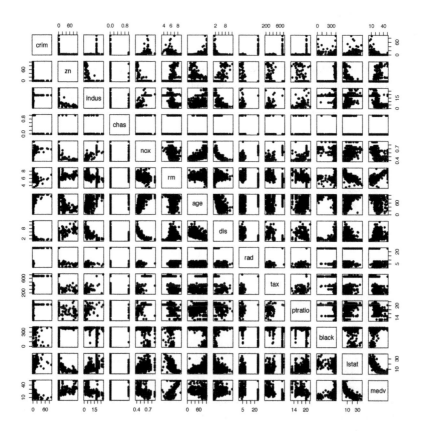

FIGURE 8.5: A scatterplot matrix for all fourteen variables in the *Boston* dataset.

Parallel coordinate plots

Figure 8.6 is the default parallel coordinates plot using the function `parcoord` from the package **MASS**. Some of the features can be seen that were identified in the scatterplot matrix display, at least for those variables with adjacent axes. There is quite a lot of information on the distributions of the individual variables, such as the skewness of `crim` and the gaps in `rad`, `tax`, and possibly `black`.

Parallel coordinates are most effective used interactively, when groups of points can be selected across all axes and be compared with the rest. This can be done in the package **iplots** and its possible successor, the package **Acinonyx**, which is in development. Figure 6.19 shows the same plot, giving some of the flavour of interaction, with the points where `rad=` 24 highlighted in blue and with the variable axes ordered by differences between those cases and the rest.

```
data(Boston, package="MASS")
par(mfrow=c(1,1), mar=c(2.1, 1.1, 1.1, 1.1))
MASS::parcoord(Boston)
```

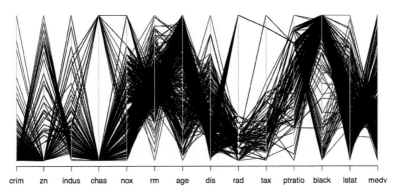

FIGURE 8.6: A parallel coordinates plot for all fourteen variables in the *Boston* dataset. Details are discussed in the text.

Heatmaps

With a heatmap each case is represented by a row and each variable by a column. The individual cells are coloured according to the case value relative to the other values in the column. For this purpose the variables are standardised individually, either with a normal transformation to z scores or adjusted to a scale from minimum to maximum. (It is possible to colour according to all values in the dataset, although that is unwise, as it emphasises differences between the levels of different variables rather than differences between individual cases.)

As the orderings of cases and variables may be freely chosen, it is helpful to try to order them in an informative manner. Clustering or seriation methods can work

```
library(gplots)
heatmap.2(as.matrix(Boston), scale="column", trace="none")
```

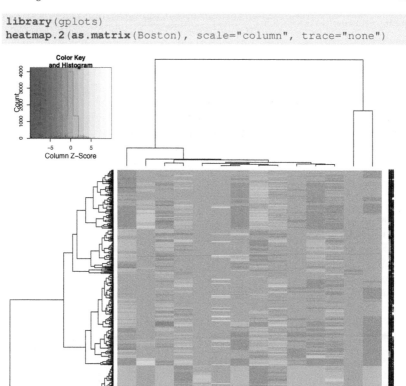

FIGURE 8.7: A heatmap for the *Boston* dataset. Some features can be discerned with a little effort (see text).

well and you just have to bear in mind that each method will give different results. Figure 8.7 shows a heatmap of the Boston data using the package **gplots**.

It is difficult to see much, although certain patterns are apparent, such as the group of relatively high values for the variable `black` because of the shape of the distribution, and the blocks of equally shaded values on some variables in the lower section of the plot. It is difficult to see much more. The colour legend top left with the superimposed histogram of values is useful, because we can see that although there are many different possible shades, most of the data values lie in the centre of the scale and only a few shades have been used. Using a different colour palette and possibly a nonlinear scale could make the display more enlightening. Experimenting with various colour schemes and clustering methods might reveal additional information.

All this makes heatmaps a fairly subjective tool and it is one of those graphic displays which can be effective for particular structures in some datasets, but which cannot be relied upon to produce good results in general.

Glyphs

With glyphs each case is represented by a multivariate symbol reflecting the case's variable values. As for heatmaps, each variable must be standardised in some way first and this can influence the way the display looks a lot. The type of symbol used is also relevant and makes a big difference. It could be the oft discussed Chernoff faces, profile charts, star shapes, or some other form. Whatever is used must have at least the same number of dimensions as the number of variables in the dataset and each variable is allocated to one (or more) dimensions.

In Figure 8.8 only glyphs for the first four cases have been drawn to show some of the details of the plot. The stars function has been used and the segments have been coloured for better effect using a rainbow palette. You really need a big screen or zooming capability to appreciate the display of the full dataset.

```
par(mar=c(1.1, 1.1, 1.1, 1.1))
palette(rainbow(14, s = 0.6, v = 0.75))
stars(Boston[1:4,], labels=NULL, draw.segments = TRUE)
```

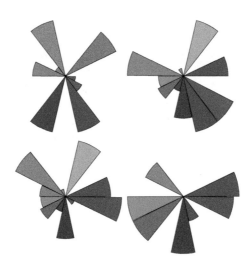

FIGURE 8.8: Glyphs (stars) for the first four cases of the *Boston* dataset.

Figure 8.9 shows the result for the whole Boston dataset. It looks like that there are several distinct groups in the dataset, as we can see groups of different shapes. Surprisingly the data show some evidence of grouping already. As with heatmaps, the allocation of the variables to the dimensions (in Figure 8.9 this is the ordering of the variables round the star), the scales used, and the ordering of the cases can strongly influence what information can be detected.

```
stars(Boston, labels=NULL, draw.segments = TRUE)
```

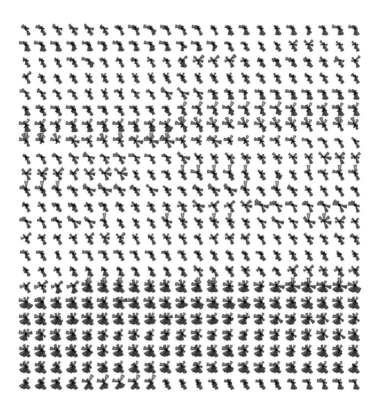

FIGURE 8.9: Glyphs (stars) for the *Boston* dataset. The bigger glyphs towards the foot of the plot are the cases where the variable `rad` takes the value 24.

8.4   Multivariate overviews for categorical variables

All the displays in the previous section are primarily for continuous variables, although they can be useful for categorical variables sometimes too. If you want to look at a small group of categorical variables together, then some kind of mosaicplot is best. This can be useful in checking experiments to see whether a study is unbalanced, and, if so, how.

The famous *barley* dataset, which Cleveland reanalysed in his book [Cleveland, 1993], has three categorical variables and one yield measurement for each combination. You can immediately see that the experiment is balanced by drawing a mosaicplot of the categorical variables and observing the resulting regular pattern.

The dataset *foster* in the package **HSAUR2** has two categorical variables, the mother's genotype and a litter genotype. Figure 8.10 shows that the structure is unbalanced, as the rectangles representing the variable combinations have different sizes.

```
data(foster, package="HSAUR2")
mosaic(~litgen+motgen, data=foster)
```

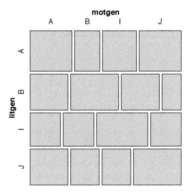

FIGURE 8.10: Genotype groups in the *foster* dataset using a mosaicplot. There are not equal numbers in the different combinations.

It is more informative to use a multiple barcharts version of a mosaicplot, which can be drawn using **ggplot2**'s functionality, as in Figure 8.11. Then it is easier to see just which groups are smaller or bigger than average.

As was mentioned in Chapter 7, there are distinct limits to the numbers of categorical variable combinations that can reasonably be displayed and understood. This is fine for monitoring experimental designs, as experiments usually only have a restricted number of combinations. It can be an issue in large surveys where there may be many classifying variables to be taken into account. The dataset *HI* in **Ecdat**

```
ggplot(data=foster, aes(motgen)) + geom_bar() +
    facet_grid(litgen~ .) + xlab("") + ylab("") +
    scale_y_continuous(breaks=seq(0,6,3)) +
    labs(title="litter genotype by mother's genotype")
```

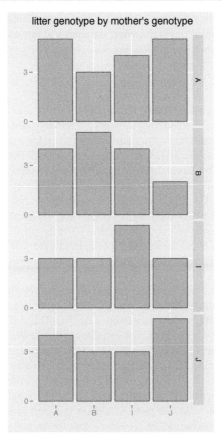

FIGURE 8.11: Genotype groups in the *foster* dataset using a multiple barchart. Numbers of rats vary across the different genotype combinations.

includes information for 22,272 married women on region (4 categories), race (3), education (6), as well as three variables on insurance status amounting to 6 different categorisations. This gives 432 combinations in total that might be of interest. There is also a weighting for each case.

8.5   Graphics by group

Sometimes there are known groups in a dataset and it is important to get an overview of variable values split by group. There are two typical situations that arise: you can have several grouping or conditioning variables and just a couple of variables to display or you can have a single grouping variable and many variables to display.

Trellis graphics

Trellis graphics [Becker et al., 1996] are ideal for the first case. They were introduced some twenty years ago to effectively display data for large numbers of subsets. Each component plot or panel shows the same basic display for a different subset of the data, but each has the same scaling to allow comparison. Subsets can be defined by categorical variables, by discretisations of continuous variables and by combinations of variables. Each of the individual panels is a conditional view of the data.

In R you can use the packages **lattice** or **ggplot2** for this. There was an excellent comparison of the two approaches on the blog Learning R [rlearnr, 2009] in 2009. Trellis graphics can be very effective and some people use them a lot. There is extensive information about the **lattice** package in [Sarkar, 2008] and on the accompanying webpage. Information on **ggplot2** is available in [Wickham, 2009] and on the **ggplot2** webpages.

Figure 8.12 shows a lattice display for the *barley* dataset. Apart from using filled circles for the points rather than open circles, this plot just uses the graphics defaults. It shows that the increasing yields across the sites from Grand Rapids to Waseca hold for all varieties except for Peatland. You can also verify Cleveland's observation, that the 1932 yields are almost always less than the 1931 yields, except for Morris where it is the other way round. This is easier to see with Cleveland's plot, where there are six panel plots, one for each site, rather than this plot where there are ten panel plots, one for each variety. On the other hand, the different pattern for the Peatland variety is easier to see with this plot. As always it is best to look at a selection of graphics. Cleveland concluded that the Morris data for the two years must have been switched. Recently Wright has re-examined the dataset using additional sources of supplementary data and suggests that in the light of the variability in the data he has found, the Morris data are quite plausible as they are [Wright, 2013].

Trellis graphics may be drawn in many different ways, depending on the choice of panel variables and panel plot, depending on the conditioning variables and what order they are in, depending on the order of categories within a conditioning variable (**lattice** plots the sites and varieties for *barley* in increasing order of median yield, as suggested by Cleveland, because that is how the factors are ordered in the R dataset), and depending on how the individual plots are arranged on the page. How well the plots look and how informative they are also depends very much on the size and aspect ratio of the overall display. Draw Figure 8.12 yourself and experiment with growing and shrinking the window in both directions.

```
library(lattice)
data(barley, package="lattice")
dotplot(site ~ yield |variety , data = barley,
        groups = year, columns=2, pch=16, col=c("red","blue"),
        key = list(text=list(levels(barley$year)),
        points = list(pch=16, col=c("red", "blue"))),
        xlab = "Barley Yield (bushels/acre) ", ylab=NULL,
        main="Barley Yields by Site for ten Varieties")
```

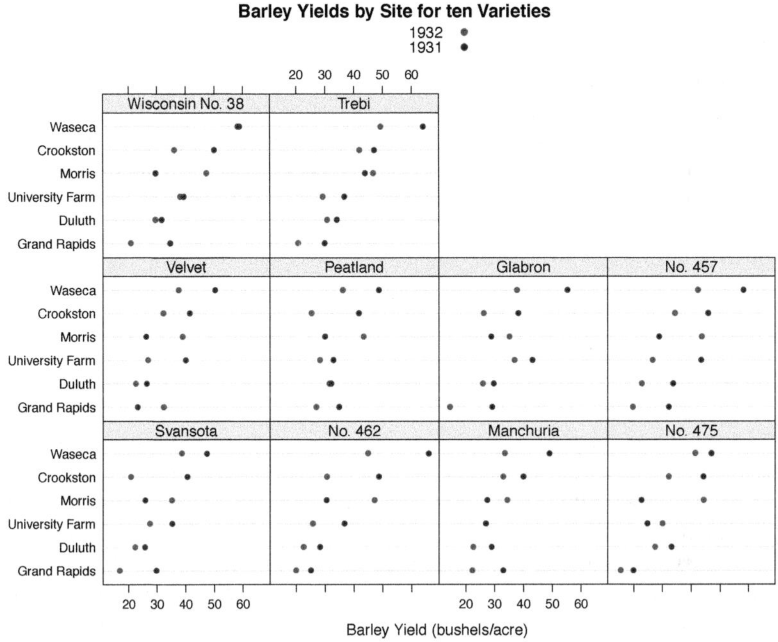

FIGURE 8.12: A lattice plot of the *barley* data. Yields at six different sites for two years are shown in ten plots, one for each variety. Yields in 1932 were generally lower than in 1931 and there is a common pattern across sites for most varieties.

Like mosaicplots, trellis graphics can in theory include unlimited numbers of combinations, in practice the individual plots become too small if you try to get everything on one page. When trellis graphics were first introduced, applications were described using hundreds of plots printed on many, many pages. This is a sensible approach if you are looking for individual plots which stand out, although it is difficult to get an overview and an idea of overall structure. For designed experiments and other structured datasets it will often be possible to organise all plots on one page, as the number of combinations usually remains limited.

Group plots

When there is only one grouping variable, it is interesting to look at how all other variables vary in parallel and for this a group plot can be drawn. Either there is a column for each variable to be displayed and a row for each group or the other way round. Mostly columns are better for comparisons, unless boxplots are used. Continuous variables may be plotted as histograms (density estimates or other displays could be used), whereas categorical and ordinal variables may be plotted as barcharts. In group plots all plots for the same variables are drawn to the same scale to aid comparison.

Figure 8.13 shows an example for the *uniranks* dataset for UK universities from the **GDAdata** package, which was discussed in §6.6. Histograms have been drawn for all the nine variables for each of the six groups of universities. Rather than loop through the variables, the code constructs a long version of the dataset and then uses facetting to arrange the plots. The `scales = "free_x"` option allows each column to have its own x-axis scale.

Several interesting features can be seen: the top performance of the Russell group, the good performance of the 1994 group, the range of performances for the universities which do not belong to any group, and the roughly equally poor showing of the other three groups. Inspecting individual columns more closely, you can see the sharp divisions on `EntryTariff` between the groups and to a lesser extent a similar effect for `AvTeachScore`.

Both trellis graphics and group plots are made up of lots of individual plots. With trellis displays every individual plot is of the same type, shows the same variables, and has the same scale, they just each show a different subset of the data. With group plots each column is like a trellis plot, but different columns show different variables and have different scales while each row shows the same subset. For all continuous variables or all categorical variables each row is like a set of histograms or barcharts respectively, treating that subset as a dataset in its own right, as described in Section 8.2. The only differences are that the plots are drawn in one row and the scaling of the individual plots depends on the scaling of the other plots in the same column for the other subsets.

Group plots are not seen very often, which is perhaps surprising. They are simple to understand and they offer effective multivariate comparisons. Since their main advantage lies in comparisons down columns there is no drawback in drawing several of them if there are many variables involved. Figure 8.13 is just about large enough to include all nine variables. Had there been more variables, then they could have been drawn in several displays over more than one page.

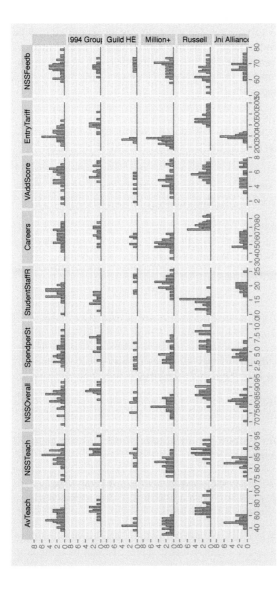

FIGURE 8.13: A group plot of the *uniranks* dataset. The Russell universities tend to get top scores on all criteria (it is better to have low values for `StudentStaffRatio`) except for `NSSFeedback`. The top row representing universities who are not members of any group have a spread of values on all criteria. The 1994 universities look to be second best to the Russell universities.

```
data(uniranks, package="GDAdata")
names(uniranks)[c(5, 6, 8, 9, 10, 11, 13)] <- c("AvTeach", "NSSTeach", "SpendperSt", "StudentStaffR",
                                                 "Careers", "VAddScore", "NSSFeedb")

ur2 <- melt(uniranks[, c(3, 5:13)], id.vars="UniGroup", variable.name="uniV", value.name="uniX")
ggplot(ur2, aes(uniX)) + geom_histogram() + xlab("") + ylab("") +
    facet_grid(UniGroup~uniV, scales = "free_x")
```

8.6   Modelling and testing for overviews

1. Transformations
   If a first look shows that distributions of variables are highly skewed, it may be
   constructive to transform the variables. The Box-Cox family of transformations
   is a good choice. If nominal variables have many categories with small frequencies, it may be helpful to combine or delete some. Otherwise, tests mentioned in
   Chapters 3 and 4 may be appropriate.

2. Associations (and causality)
   When variables show associations in a scatterplot matrix, then linear or nonlinear models, depending on the form of association, should be considered.

3. Linear models
   If one or more panel plots stand out as being different in a trellis display, then a
   linear model based on the conditioning factors might be used to confirm this.

4. Discrimination between groups
   Discriminant analysis could be used in conjunction with group plots, as could
   other supervised learning methods, including Support Vector Machines.

Main points

1. Sets of univariate graphics are good for giving a quick overview of variable values and distribution patterns (e.g., Figures 8.1 and 8.2).

2. Scatterplot matrices are valuable for identifying bivariate patterns even with quite
   a few variables (cf. Figure 8.5).

3. Parallel coordinate plots are useful for studying groups of cases and are most
   effective when they are interactive (§8.3).

4. Trellis displays are excellent for comparing data on one or two variables by subsets, while group plots provide a useful overview across many variables for a
   small set of subgroups (cf. Figures 8.12 and 8.13).

5. Other multivariate displays look interesting, but scaling, colours and orderings
   all have to be suitably chosen and the resulting graphics can still be difficult to
   decode (§8.3).

Exercises

1. **Cloud seeding** (*clouds* from the package **HSAUR2**)
   What kinds of variables are there in this dataset? What plots would you recommend to help people get to know the dataset?

2. **Longley's Data** (*longley* from the package **datasets**)
   Longley's dataset is well known as an example for highly collinear regression. Can you see this from the scatterplot matrix? Are there other features worth noting?

3. **US States** (*state.x77* from the package **datasets**)
   What kinds of variable are there? Is there anything interesting or unusual in the univariate distributions? Compare what information you might get from each of the multivariate displays discussed in this chapter: a scatterplot matrix, a parallel coordinates plot, a heatmap, and a collection of glyphs.

4. **Crabs** (*crabs* from the package **MASS**)
   [Venables and Ripley, 2002] uses this dataset in discussing classification and discrimination. The authors initially transform to a log scale and then write that "The data are very highly correlated and scatterplot matrices and brush plots [i.e. interactive graphics] are none too revealing." Using graphical overviews, comment on whether the log transform was a good idea and whether you agree with their statement on the correlations.

5. **Pima Indians**
   Data about diabetes amongst adult females of the Pima Indians is available in R, in the packages **MASS** and **MMST**. Both use the 532 cases with complete records. A larger version of the dataset with 768 cases is available from the UCI Machine Learning Repository [Bache and Lichman, 2013]. Download the larger dataset and give an overview of the differences between the cases available in R and the rest using two groups plots, one for the continuous variables and one for the variables `npreg` and `type`.

6. **Exam results in London**
   The dataset *Exam* in **mlmRev** is used to illustrate multilevel modelling. Prepare an initial graphical summary of the data and summarise your results in three main conclusions.

7. **Distance to college**
   The dataset *CollegeDistance* in **AER** is from a survey of high school students. If you had to prepare a one-page summary of the main information you can find by exploring the data, what graphics would you use?

8. **US traffic fatalities**

   Numbers of various kinds of traffic fatalities for each of six years are given for
   each state in the contiguous United States in the dataset *Fatalities* from the package **AER**. There are 32 variable values for each state and year. Carry out an
   initial exploratory analysis to decide what information you would present to give
   people a first impression of the data.

9. **Intermission**

   The painting *Britain at Play* by L.S. Lowry hangs in the *Usher Gallery* in Lin-
   coln. Does the title match the picture well? Can you see how the British 'played'
   in the early 1940s?

9

Graphics and Data Quality: How Good Are
the Data?

Beauty is less important than quality.

Eugene Ormandy

Summary

Chapter 9 discusses data quality, including missing values and outliers, what they
are, and ways of identifying them.

9.1 Introduction

Good analyses depend on good data. Checking and cleaning data are basic tasks that
always have to be carried out and yet are rarely explicitly discussed. Textbooks—and
indeed R packages—often present datasets only after they have been cleaned and
filtered. The detailed work that has gone into getting them into their semi-pristine
condition is passed over and sometimes even swept under the carpet. One of the
problems is that there are so many different ways that data can be of poor quality, or
as Hadley Wickham put it in a contribution to R Help, paraphrasing Tolstoy: "every
messy data is messy in its own way".

There are some general principles, some techniques which are useful fairly often,
and a whole raft of special cases. Graphics are helpful for identifying these problems
and can sometimes shed light on how they can be solved. Even if you cannot solve a
problem, it is important to know how serious it is.

Data quality problems may include multiple codings for the same category (e.g.,
Male, male, m, M, man, . . .), measurement or data entry errors, data heaping, gaps in
the data, ambiguous definitions of variables, shifted chunks of data, and, of course,
missing values and outliers, which are both discussed in this chapter.

177

9.2   Missing values

Missing values in datasets can lead a rather shadowy existence. If there are very few of them, they may be just ignored. If there are more of them, then values may be imputed to replace them. In either case reported results may not make clear that there were any missing values in the dataset. Political opinion polls generally report party support for the people who said they would vote for a particular party. The 'don't knows' may only be referred to in the small print, if at all, and can be a sizeable proportion of the electorate.

Visualising patterns of missing values

Graphical displays can assist in summarising how many missings there are and in ascertaining if there are patterns of missingness. In statistics textbooks there tend to be few dataset examples including missing values, possibly because they add irrelevant difficulties that may distract from the theory being illustrated in the examples. Real datasets often have missings of one kind or another.

In the R package **mi** for multiple imputation by Bayesian methods there is a dataset *CHAIN*, used earlier in Exercise 6 of Chapter 3. Figure 9.1 shows the patterns of missings using the display provided in the package.

```
data(CHAIN, package="mi")
par(mar=c(1.1, 4.1, 1.1, 2.1))
mi::missing.pattern.plot(CHAIN, y.order=TRUE, xlab="", main="")
```

FIGURE 9.1: A missing pattern plot of the *CHAIN* dataset in the package **mi**. The rows represent the seven variables and the columns the 532 cases. Where a value is missing on a variable for a case the corresponding cell is marked in red. The variables have been ordered by numbers of missings and the cases clustered. All variables have missing values for a few cases.

We can immediately see the curious fact that a number of cases have no values, just missings (the group on the left), and that only three variables have other missing values, since only those rows have other red cells. As this display attempts to show all entries for all cases individually, it is only feasible for smaller datasets.

With the function `visna` in the package **extracat**, the different missing value patterns are displayed not the individual missing values. Figure 9.2 uses the same data as in Figure 9.1 and additionally provides information on the proportions of missings by variable and the relative frequency of each pattern. The figure also emphasises that the cases with only the first variable missing make up the majority of the cases with missings.

```
visna(CHAIN, sort="b")
```

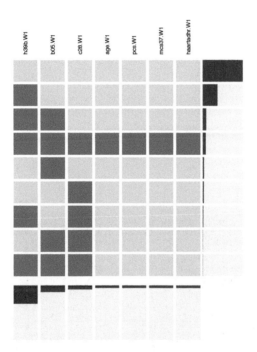

FIGURE 9.2: A missing pattern plot of the *CHAIN* dataset using the package **extracat**. The columns represent the seven variables and the rows the missing patterns. The cells for the variables with missing values in a pattern are drawn in blue. The variables and patterns have been ordered by numbers of missings on both rows and columns (`sort="b"`). The bars beneath the columns show the proportions of missings by variable and the bars on the right show the relative frequencies of the patterns. One variable has far more missings than the others. Four variables are only missing for cases where all variables are missing.

For some datasets the number of cases with no values missing may be very large. The frequency bar for this pattern is then reduced in size to enable readers to distinguish differences in the frequencies of the other patterns and the bar is given a red border to indicate this censoring.

In the *CHAIN* dataset there are seven variables, all of which have some values missing. Potentially there could have been 128 $(= 2^7)$ missing patterns, yet there are only nine. This often happens and it makes the visna plot very efficient. In the *oly12* dataset in **VGAMdata**, which contains details of the athletes who took part in the 2012 London Olympics, there are 14 variables with 3 having missing values. All eight possible patterns of missings amongst these 3 variables arise, as Figure 9.3 shows. (This dataset was already discussed in §5.5.) The large number of date of birth cases missing obscures the patterns involving weight and height. Redrawing the plot without the variable date of birth, i.e., using the subset

```
oly12d <- oly12[, names(oly12) != "DOB"]
```

shows that weight is often missing on its own, while height is mostly missing together with weight and rarely on its own.

```
data(oly12, package="VGAMdata")
oly12a <- oly12
names(oly12a) <- abbreviate(names(oly12), 3)
visna(oly12a, sort="b")
```

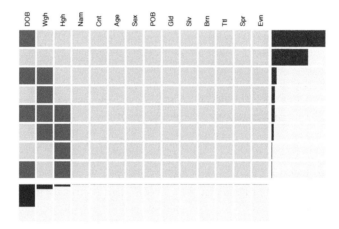

FIGURE 9.3: Missing value patterns in the London Olympics dataset (*oly12*). Date of birth is missing very often and weight is more often missing than height.

Missings dependent on values of other variables (MAR)

To investigate whether missings may be MAR, rather than MCAR, you can compare the subsets of the cases with and without missings. The dataset *freetrade* in the
Amelia package provides values of 8 variables for nine Asian countries for each of
19 years. Five of the variables have missing values, one of which being the country's
average tariff rates, where about a third of the data are missing. Figure 9.4 displays
the missing pattern for this variable by year and by country. The countries have been
sorted by the numbers of years for which the data are missing and the table of missings has been visualised with a fluctuation diagram. The tariff rates are missing more
often for some countries than for others and there is little evidence of any pattern
over time. The data do not look MCAR.

```
data(freetrade, package="Amelia")
freetrade <- within(freetrade, land1 <-
        reorder(country, tariff, function(x) sum(is.na(x))))
fluctile(xtabs(is.na(tariff) ~ land1 + year, data=freetrade))
```

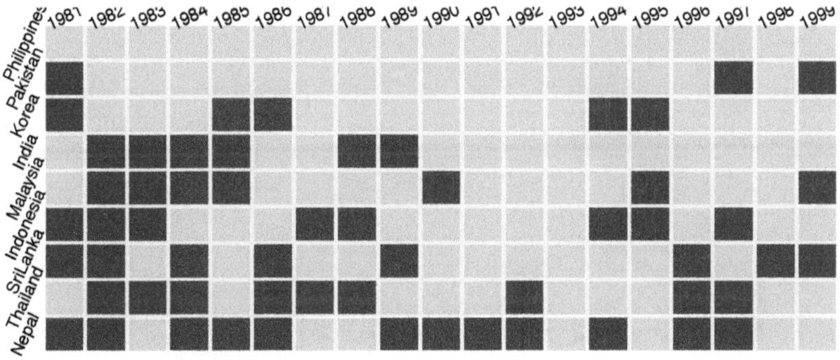

FIGURE 9.4: A plot of missing values on tariff rate information for nine Asian countries over 19 years. There is complete data for the Philippines and data for all countries in 1993. There are no striking patterns.

Reasons for missings and dealing with missings

Data can be missing for different reasons. It could be that a value was not recorded,
or that it was, but was obviously an error and was consequently coded as missing.
A value may be lost or incorrectly transcribed. In a survey people may not answer
because they don't know, because they don't want to answer, because there is no
option that matches their views, or because they were not asked. Or they may give a
ridiculous answer, which the interviewer then marks as 'missing'.

There is no standard missing code used by software (the NA used by R and other software is common). In the past, numbers like 99 or 999 were used and caused problems when people didn't notice and calculated statistics and fitted models as if those were real numbers. The version of the Pima Indians dataset offered on the UCI Machine Learning Library uses 0 for entries the R version records as missings. Figure 9.5 shows a missing pattern plot for the *Pima.tr2* dataset from **MASS**. Most of the records are complete. Quite a few cases are only missing a skin thickness value and a few values are missing on two of the other variables.

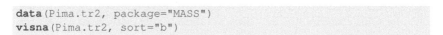

```
data(Pima.tr2, package="MASS")
visna(Pima.tr2, sort="b")
```

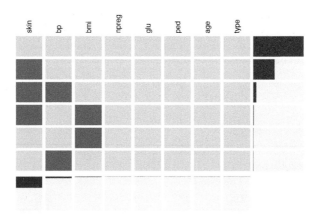

FIGURE 9.5: A missing value pattern plot for the Pima Indians dataset. Only three variables have missing values and no case is missing on all three.

The best that can be said is that all data (and documentation) should be looked at carefully before analysis. Other major statistics packages (SAS, SPSS, Stata) offer multiple missing codings to take account of different reasons for data being missing, but R does not; codings for more than one type of missing have to be sorted out by the user. As an example in practice, the US Multicenter Osteoarthritis Study (http://most.ucsf.edu/) has 13 different missing codes in its protocol.

Taking account of missing values can be important, especially if they are 'not missing at random' (NMAR). This is a term that became popular when multiple imputation was introduced [Rubin, 1987], the clever idea that instead of imputing a single value to substitute for a missing value, you should use several. NMAR data are distinguished from MCAR (Missing Completely At Random) and MAR (Missing At Random) (where missingness may depend on the value of the missing entry). One of the reasons the 1992 election in the UK was badly predicted was apparently that conservative voters were reluctant to admit to intending to vote conservative [Worcester, 1996], an example of MAR.

9.3 Outliers

What is an outlier?

Outliers are cases which are far away from the bulk of the data. They may be errors (a decimal point in the wrong place), genuine extreme values (some people are paid much more money than the rest of us), rare values (occasionally a lot of people are sick on the same day), unusual values (small people can be very heavy), cases of special interest (exceptional performances), or data from some other source (including a few small companies in a dataset on large companies). The last is sometimes referred to in the statistical literature as contamination.

Whether points are judged to be outliers depends on the data or model with which they are compared. A basketball player may be regarded as unusually tall amongst a group of his own age, whilst being average within a group of his fellow basketball players. Outliers are often talked about as if they are individual unusual values, but you often observe groups of outliers, particularly with large datasets, and one analyst's outlying group may be another person's population subset.

As groups of outliers get larger relative to the main part of a dataset, you could say you are looking for outliers or you are identifying distinct subsets or you are assessing how homogeneous the dataset is. This chapter concentrates on at most small groups of outliers and does not consider the more complex problem of dataset heterogeneity in large datasets. That is a topic for itself and depends particularly heavily on the context and aims of a study.

Possible outliers on individual variables are easy to spot graphically, and bivariate outliers, which are not outlying on the individual variables, can be seen in scatterplots. There can also be higher dimensional outliers that are not outliers on lower numbers of dimensions and these are trickier to find. That is hardly surprising: Try to think of what a three-dimensional outlier looks like which is neither a one-dimensional outlier on any of its three dimensions nor a bivariate outlier on any of the three pairs of dimensions.

Determining outliers analytically is not easy. Many procedures have been suggested, but none are universally accepted, there are too many different kinds of outlier that might arise. Analytic approaches are useful for picking out extreme outliers (and every method should be able to manage that) and for providing a first filtering of the data to produce a set of possible outliers that can then be examined in more detail.

It is worth identifying outliers for a number of reasons. Bad errors should be corrected, genuine outlying values can be interesting in their own right, and many statistical methods may work poorly in the presence of outliers. Robust methods have been proposed for dealing with datasets with outliers. They are often computationally demanding.

There is a fairly extensive literature on outliers beginning with the classic text by Barnett and Lewis [Barnett and Lewis, 1994] and including more recently [Aggarwal, 2013].

Examples of outliers

Figure 9.6 shows a sample collection of plots for identifying potential outliers in a subset of the *USCrimes* dataset from the package **TeachingDemos**. You can see a boxplot of the murder rates in US States in 2010, a scatterplot of the rates of theft and vehicle theft, and a parallel coordinate plot of the rates for nine different crimes.

The preparatory code for the plot involves some subsetting and restructuring. One important point to note is that the dataset *USCrimes* is a three-dimensional array, not a data frame, so it has to be converted to a table and a long data frame first. Then data for the year 2010 are chosen and the data for the US as a whole are excluded. Next the data frame is reformed with columns for each of the variables using the `spread` function from **tidyr**. Finally the rate variables are selected:

```
library(tidyr)
data(USCrimes, package="TeachingDemos")
names(dimnames(USCrimes)) <- c("State", "Year", "Crime")
US10 <- USCrimes %>% as.table %>%
        as.data.frame(responseName = "Rate") %>%
        filter(Year==2010 & State != "United States-Total") %>%
        spread(Crime, Rate) %>% select(State, ends_with("Rate"))
```

The boxplot identifies two possible outliers and one is clearly hugely different from all the rest, the District of Columbia. This can be identified with

```
US10 %>% select(State, MurderRate) %>% filter(MurderRate > 10)
```

The scatterplot seems to show one extreme outlier on both theft rates, the District of Columbia again. Interestingly, a boxplot of `TheftRate` alone shows no outliers. The scatterplot also indicates two states looking like bivariate outliers. Neither California nor Nevada are extreme on the individual variables, but both have relatively higher vehicle theft rates compared to their theft rates. Finally the parallel coordinate plot, in which each axis has been scaled individually between 0 and 1, reveals that one state has extremely high values on six of the nine variables, obviously the District of Columbia and that another state has a particularly high rate of rape.

```
a <- ggplot(US10, aes("var", MurderRate)) +
            geom_boxplot() + xlab("") +
            ylab("Murder rate per 100,000 population")
b <- ggplot(US10, aes(TheftRate, VehicleTheftRate)) +
            geom_point() +
            xlab("Theft rate per 100,000 population") +
            ylab("Vehicle theft rate per 100,000 population")
c <- ggparcoord(data = US10, columns = c(2:10),
                scale="uniminmax") +
                theme(axis.title.x = element_blank(),
                axis.title.y = element_blank())
grid.arrange(arrangeGrob(a, b, ncol=2, widths=c(1, 4)),
                c, nrow=2)
```

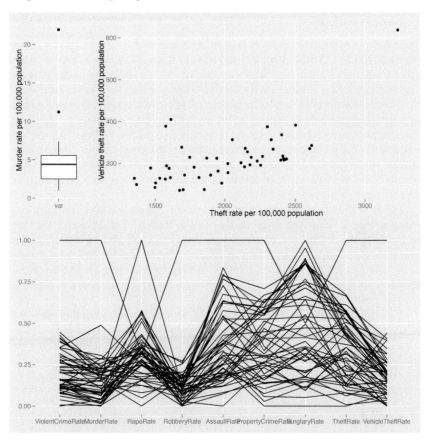

FIGURE 9.6: Plots of the crime rates by state in 2010 for the *USCrimes* dataset, used for identifying potential outliers.

Univariate outliers

There have been many attempts to suggest rules and tests for determining whether points should be considered outliers and none of them are satisfactory in all situations. Using tests for assessing whether points are outliers is difficult, as the tests are primarily designed for testing individual points and you may want to test several.

The best-known approach for an initial look at the data is to use boxplots. Tukey suggested marking individual cases as outliers if they were more than $1.5 \times IQR$ (the interquartile range) outside the hinges (basically the quartiles). Figure 1.8 showed boxplots for the six continuous variables in the Pima Indians dataset. There are no outliers on the first variable (plasma glucose concentration), both low and high outliers on blood pressure, and high outliers on the other four variables, with as many as nine for the diabetes pedigree function.

Figure 9.7 displays boxplots for the seven continuous variables from the *diamonds* dataset from package **ggplot2**, also used in Exercise 7 of Chapter 3. First the code selects the desired variables and then combines them all in a new long data frame, so that all plots can be drawn simultaneously with the last line. Note the use of the `scales` option to ensure individual scaling for each variable and the `nrow` option to put all plots in one row. The formatting options are necessary to clear the plots of irrelevant labelling.

The differences in how Figures 1.8 and 9.7 were drawn and what they show are worth noting. Traditional base **graphics** were used for Figure 1.8 and the boxplots for all the variables are shown in one display. This required that the variables were transformed to a common scale first. Graphics from **ggplot2** were used for Figure 9.7. By pretending that all the variables were actually one long variable composed of different subsets it was possible to draw individual plots for each of them with a single command, using their actual scales.

Boxplots generally work well, although they tend to imply too many outliers for large datasets such as *diamonds*. The reason is clear: With large datasets there are sufficient numbers of points at intermediate outlying levels for them to be regarded as part of the population, even though boxplots plot them as outliers. At any rate, you can argue that it is better to identify too many potential outliers as too few. You are probably going to review the outliers in their order of 'outlyingness' anyway, and so the more extreme ones will be dealt with first.

Apart from obvious errors (e.g., digits have been transcribed, so that a temperature of 19° has been written as 91°), you need background knowledge to judge what kind of case an indicated outlier might be. In most applications there is information on many other variables and this can be used to help decide. Scatterplots can be used for looking at pairs of variables (as in Figure 9.8), while parallel coordinate plots are good for looking at more. Perhaps the case is an outlier on several variables, either in a consistent way (the District of Columbia in the *USCrimes* dataset in Figure 9.6) or in an inconsistent way (sometimes big, sometimes small). The latter may suggest a data shift error, where values have been entered in the wrong columns for a case.

When data have skew distributions, it is often a good idea to transform them, perhaps to a logarithmic scale. Points which were high outliers on the original scale

```
library(tidyr)
diam1 <- diamonds %>% select(carat, depth:z) %>%
    gather(dX, dV, carat:z)
ggplot(diam1, aes("dX", dV)) + geom_boxplot() +
    facet_wrap(~dX, scales = "free_y", nrow=1) +
    xlab("") + ylab("") + scale_x_discrete(breaks=NULL)
```

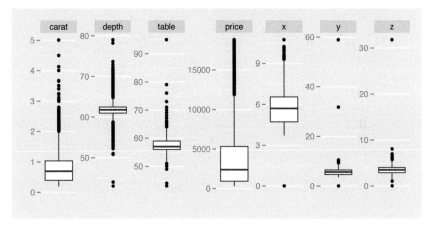

FIGURE 9.7: Individual boxplots of seven of the *diamonds* variables. Each of the variables has outliers, which is hardly surprising in a dataset of over 50 thousand cases. The last three variables, x, y, and z have cases with value 0, which must be wrong, and there are several high outliers, which must be wrong too. Both the carat and price variables are skewed to the right.

will then no longer necessarily be outliers on the new scale and some points that were formerly regarded as non-outliers may now be low outliers. Again, background knowledge is essential to judge whether a transformation makes sense.

Multivariate outliers

Most outliers are outliers on at least one univariate dimension, sometimes several. The more variables there are, the more likely it is that cases are outliers on some dimension or other. This implies that the more variables you have, the more stringent the rules should be for deciding which cases might be outliers. In practice you usually start with the most extreme outliers and work inwards.

Multivariate measures depend on robust estimation of the parameters of the data. For small datasets the efficiency of calculating these robust statistics is good and there can only be a few outliers anyway. At least, you would think so. The famous *stackloss* dataset, a favourite amongst robust statisticians, has only 21 points and probably all of them have been declared an outlier at one time or another in some article or other [Dodge, 1996].

Large datasets are more complicated. It is much more likely that they are heterogeneous, the data quality is probably poorer, and there may well be a variety of
outliers and outlying groups of diverse kinds. Whether this has much influence on
an analysis is another matter. The *diamonds* dataset shows an example. The first plot
in Figure 9.8 is a scatterplot of y and z (actually width and depth). There are some
obvious errors (values of 0 and excessively high values). Restricting the ranges gives
the second plot, which excludes 23 of the almost 54,000 points. Now some other
cases appear to be in error (the three low values on depth), while others look unusual
compared to the bulk of the data, bivariate outliers that are not outlying on either of
width or depth alone.

```
a2 <- ggplot(diamonds, aes(y, z)) + geom_point() +
            xlab("width") + ylab("depth")
d2 <- filter(diamonds, y > 2 & y < 11 & z > 1 & z < 7)
b2 <- ggplot(d2, aes(y, z)) + geom_point() +
            xlab("width") + ylab("depth")
grid.arrange(a2, b2, ncol=2)
```

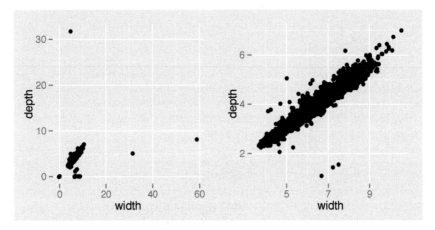

FIGURE 9.8: Plots of *y* (width) and *z* (depth) from the *diamonds* dataset with and
without 23 outliers and rescaled accordingly. Without the extreme outliers the plot is
more informative and additional cases look suspicious.

The statistics of the dataset including and excluding the original 23 outliers barely
change, only the graphics are dramatically affected and that is because of the univariate outliers. Note how Hadley Wickham's **dplyr** package is used for calculating the
statistics.

```
diamS1 <- summarise(diamonds, meanx = mean(x), sdx = sd(x),
                meany = mean(y), sdy = sd(y), corxy = cor(x,y))
diamS2 <- summarise(d2, meanx = mean(x), sdx = sd(x),
                meany = mean(y), sdy = sd(y), corxy = cor(x,y))
```

```
diamS1

#          meanx        sdx      meany        sdy      corxy
#    1  5.731157  1.121761  5.734526  1.142135  0.9747015

diamS2

#          meanx        sdx      meany        sdy      corxy
#    1  5.731605  1.119402  5.733428  1.111272  0.9986573
```

Figure 9.8 also illustrates how points can be outliers on two dimensions without being outliers on the individual dimensions.

The decisions as to which points are outliers can be problematic and may depend on an implicitly assumed model. In Figure 9.9, showing the olive oil dataset used in two exercises in earlier chapters, you could regard points as outliers that are far from the mass of the data, or you could regard points as outliers that do not fit the smooth model well. Some points are outliers on both criteria.

```
data(olives, package="extracat")
ggplot(data=olives, aes(x=oleic, y=palmitic)) + geom_point() +
       geom_density2d(bins=4, col="red") + geom_smooth()
```

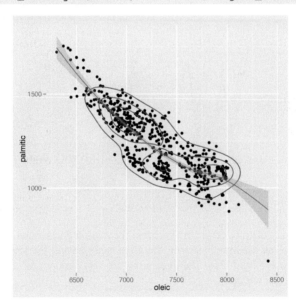

FIGURE 9.9: A scatterplot of two of the variables from the *olives* dataset with contour lines from a two-dimensional density estimate and a loess smoother superimposed. Some points to the top left are outliers according to the density estimate, but not according to the smoother.

Higher dimensional outliers are tough to spot visually when they are not extremes on lower dimensions, and it is an open question how often such cases occur in real datasets. Suggested methods of finding them include using a robustified Mahalanobis distance and there have been several proposals how to do this in the literature [Ben-Gal, 2005]. Graphical displays such as parallel coordinate plots and scatterplot matrices can then be helpful for establishing why points have been identified as outliers.

A simple illustration is given in Figure 9.10, using the *Boston* dataset yet again, where the boxplot shows that there are low outliers on the variable `ptratio`, the pupil teacher ratio. Drawing a parallel coordinate plot of all the variables and colouring the three cases sharing the lowest value gives the plot on the right.

```
data(Boston, package="MASS")
a <- ggplot(Boston, aes("var", ptratio)) + geom_boxplot() +
            xlab("") + ylab("Pupil-teacher ratio")   +
            scale_x_discrete(breaks=NULL)
Boston <- within(Boston, pt1 <- ifelse(ptratio < 13, 1, 0))
oc <- order(Boston$pt1)
b <- ggparcoord(data = Boston[oc,], columns = c(1:14),
            scale="uniminmax", groupColumn="pt1") +
            theme(axis.title.x = element_blank(),
            axis.title.y = element_blank())
grid.arrange(a, b, nrow=1, widths=c(1,4))
```

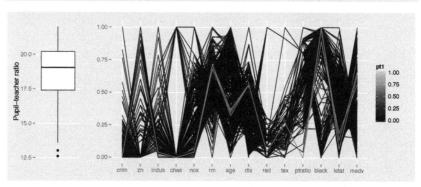

FIGURE 9.10: A boxplot of `ptratio` from the *Boston* dataset and a parallel coordinate plot of all the variables with the three cases having the lowest value for `ptratio` coloured light blue. The three areas are very similar and are not outliers on any of the other variables.

Categorical variables divide datasets into subsets and often cases can be outliers within a subgroup but not on the whole dataset or the other way round. For a single subsetting variable, the best graphical approach is to plot boxplots for each of the categories. Figure 9.11 illustrates this for `Sepal.Width` in the *iris* dataset (cf. §1.3), where the boxplot for the three species together suggests four outliers, while

```
a <- ggplot(iris, aes("boxplot for all", Sepal.Width)) +
        xlab("")  + geom_boxplot() +
        scale_x_discrete(breaks=NULL)
b <- ggplot(iris, aes(Species, Sepal.Width)) +
        geom_boxplot() + xlab("")
grid.arrange(a, b, nrow=1, widths=c(1,2))
```

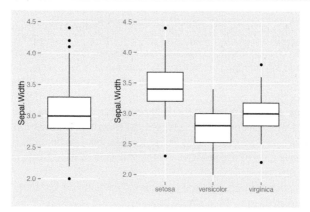

FIGURE 9.11: Boxplots of Sepal.Width for the *iris* dataset as a whole and for the three species separately. Four possible outliers are identified in each plot, while only one outlier is identified in both.

the boxplots for the three species separately also suggest four outliers, although only one of the points is an outlier in both displays.

Categorical outliers

It is easy enough to understand what is meant by an outlier on any continuous scale, but categorical outliers are rather different. The term could refer to cases which have some rare value on a categorical scale (e.g., purple hair), but that is more a matter of defining the categories appropriately for the analyses to be carried out.

The more interesting situation is where there are a number of categorical variables and certain combinations occur very infrequently or should not occur at all. In his study of data cleaning, [Malik, 2010], Waqas Ahmed Malik used a dataset from the Pakistani Labour Force Survey of 2003. He found a few cases of people who were divorced or widowed and who were classified on another variable as being the spouse of the head of household. Those cases were clearly in error on at least one of the variables. In another analysis he found people with low education and employment status recorded as being corporate managers. In a dataset with around 140,000 cases these kinds of issues are bound to arise. Fluctuation diagrams (cf. §7.4) are useful visual tools for finding such cases.

Dealing with outliers

Identifying cases as potential outliers, whether with graphical or with analytic methods, is relatively straightforward, what to do about the information is not. Outliers need to be inspected and reviewed. Sometimes they may be interesting in their own right (e.g., telephone companies have regarded unusual records as indicating fraud), sometimes they may be recognisable special cases (e.g., Bob Beamon's long jump at the 1968 Olympic Games), and sometimes they may be just errors.

If the cases are obviously wrong, then those values may be discarded or, if possible, corrected. You can decide to discard whole cases or just the unacceptable values, and background knowledge is essential for these kinds of decisions. If only individual values are discarded, they may be treated as missing and substitute values may be imputed for them in one way or another. Imputation can be a tricky business and the appropriate expertise is called for. Whenever data are discarded, it is a sensible idea to record what has been done and why, to ensure that discussion of results is complete. When two people analyse the same data, but handle outliers in different ways, it can save a lot of time to have complete details of what has been done when you compare results.

Some statistics are little affected by outliers, medians for instance, but graphics are always affected, even by a single case, as default scales have to be drawn to incorporate all the data. (Of course, this only applies to outliers which are outlying on at least one dimension, cf. Figure 9.8. Solely high dimensional outliers do not affect scaling.) One extreme value in a million can reduce the rest of the dataset to a column in a histogram or to a heavily overdrawn point in a scatterplot. To see the structure of the main part of the dataset you have to zoom in, effectively ignoring the outliers. It would be helpful if graphics gave some clue when some points are excluded like this; you can easily forget about outliers that are not visible.

Keeping outliers and discarding them are the two alternative extremes. In modelling, robust methods attempt to reduce the effect of outliers by calculating a weighting for each case. Outliers that are far away may have little or no weight attached to them, whereas cases which are neither far away nor part of the main pattern of the data may be assigned a larger weight, although still lower than cases in the middle of the data. Each robust method will produce different weightings, and the choice of parameters within a method will also have an effect. The jury is still out on how best to proceed.

A possible strategy for outliers

'Extreme' outliers are likely to be errors or data from a quite different population and should be discarded or replaced with imputed values. 'Ordinary' outliers may be of some interest in their own right. Quite often they will affect statistical modelling less than one might fear, although it is as well to check what influence they have using a robust approach, if that is possible. Initially an exploration of the dataset with graphical methods and some simple outlier rules (e.g., a robust Mahalanobis distance) should suffice to get a feel for the data.

1. Plot the one-dimensional distributions. Examine any 'extreme' outliers to see if they are interesting in their own right, if they are obvious errors which can be corrected, or if they are values which should perhaps be discarded.

2. For outliers which are 'extreme' on one dimension, examine their values on other dimensions. This can help to decide what to do about them. Consider imputing values for outlying values you intend to discard.

3. After dealing with the 'extreme' values, examine further potential one-dimensional outliers, i.e., those cases which are not 'extreme', but are marked as outliers by a boxplot or by some other rule. For these cases check the values on other dimensions using multivariate graphics, parallel coordinate plots, or sploms. Consider discarding cases which are outlying on more than one dimension.

4. Cases which are outliers in higher dimensions, but not in lower ones, should be considered next. They may be identified in sploms (bivariate outliers), or in parallel coordinate plots, but it is probably best to use a robust multivariate approach and then inspect the possible outliers graphically to see why the analysis picked them out.

5. Finally, consider whether outliers should be looked for in subsets of the complete dataset. Drawing boxplots by grouping variables is a good way of doing this. And, as always, any potential outliers found should be examined in conjunction with the case values on other variables and in the context of the study aims.

9.4   Modelling and testing for data quality

1. Missing values
   In principle some kind of binary pattern test could be carried out. In practice it is not done, probably because either there are not many missings or they have a readily interpretable pattern.

2. Univariate outliers
   Many tests have been suggested and a helpful place to start is Barnett and Lewis [Barnett and Lewis, 1994]. Some unvariate tests are implemented in the package **outliers**.

3. Multivariate outliers
   To use a criterion to identify a point as an outlier or to test for an outlier implies some kind of model of 'standard' data. No test is fully satisfactory, so much depends on the definition of 'good' data, on the possible numbers of outliers, and on the context. Different models and procedures will lead to different results.

4. Robust approaches

Modern approaches rely on robust estimation of the 'good' part of the data and, as with all robust methods, there is a wide range of possibilities involving clever heuristics of one kind or another. Robust Mahalanobis distances are key. Packages implementing the methods include **robust base**, **mvoutlier**, and **CerioliOutlierDetection**.

Main points

1. Quality of data can be a major issue and there are many ways the quality of a dataset may be poor. All manner of different approaches may be helpful in identifying and solving the problems.

2. Missing value pattern plots are good for identifying possible structure in missings (cf. Figures 9.2 and 9.3).

3. Outliers are difficult to define and pin down (§9.3). They may be compared to models or densities (Figure 9.9) or by subgroups (Figure 9.11).

4. Individual extreme values are easy to spot, groups of outliers are more difficult to determine. There can be many different reasons for outliers, sometimes they are errors, sometimes they are important special cases.

5. Graphical displays are useful for finding univariate outliers (Figure 9.7) and bivariate ones (Figure 9.8).

6. Sploms and parallel coordinate plots can be helpful for studying potential outliers (Figure 9.10), for instance ones identified by robust approaches.

Exercises

1. **Lung cancer**
   The dataset *lung* in the package **survival** has 228 patients and 10 variables. What missing value patterns are there?

2. **Academic Performance Index**
   The dataset *apipop* in the package **survey** comprises 6194 Californian schools. There are 37 variables in all.

   (a) Are the missing values missing at random or are there patterns?
   (b) Does excluding the variables with a majority of missing values change the picture much?

3. **Diamonds**
   Discuss the various kinds of outlier identified in a scatterplot of the variables carat and width from the *diamonds* dataset discussed in Figures 9.7 and 9.8.

4. **Beamon's longjump**
   The dataset *MexLJ* has the best longjumps by the 14 finalists in the 1968 Mexico Olympics. Is Bob Beamon's famous jump really so very different from those of the other competitors? Draw an appropriate plot for investigating this.

5. **Pearson heights**
   Pearsons' height data for fathers and sons were considered in §3.3 and §5.4.

   (a) Draw a scatterplot of the heights. Are there any cases you would regard as outliers?
   (b) Overlay your plot with a bivariate density estimate. Which cases do you think would be regarded as outliers under this model?

6. **Forbes2000**
   This dataset from the **HSAUR2** package includes data for the sales, assets, profits, and market value for each of 2000 companies in 2004. Are there many univariate outliers? What about bivariate outliers?

7. **Forbes2000 again**
   The three variables, sales, assets, and marketvalue, are all related to some extent to the size of the company. If they are transformed to logs, their distributions look far less skewed.

   (a) How many cases are still possibly outliers?
   (b) Given that some companies reported losses, the variable profits can only be transformed to logs by excluding companies who failed to make profits. Which additional companies might then be considered to be outliers?

8. **Wheat**

   Rothamstead is the longest running agricultural research station in the world and is probably best known to statisticians because Fisher worked there from 1919 to 1933. In 1910 a study of wheat yields was carried out there and Figure 9.12 shows the grain and straw yields. Which of the points in the scatterplot might be outliers? Why?

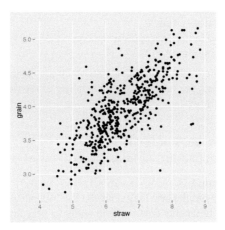

FIGURE 9.12: A scatterplot of the yields of grain and straw of 500 plots in the *mercer.wheat.uniformity* dataset from the package **agridat**.

9. **USCrimes 2010**

   Create a subset of the *USCrimes* dataset in the **TeachingDemos** package containing just the data for 2010. As the District of Columbia is an extreme outlier on several of the rates, remove it from the dataset and carry out a graphical outlier analysis.

   (a) How many univariate and bivariate outliers do you find?
   (b) Are they the same as the ones found in the dataset with the District of Columbia included?

10. **Boston housing**

    (a) Compare a boxplot and a histogram of the `tax` variable. Which is more useful for investigating outliers and how many outliers are there?
    (b) Draw a parallel coordinate plot of all the variables and colour the cases with a high value for `tax`. Are these cases different from the rest on other variables?

11. **Intermission**

    The *Mona Lisa* by Leonardo da Vinci hangs in the *Louvre* museum in Paris. Is the Mona Lisa so different from other portraits of women?

10

Comparisons, Comparisons, Comparisons

Shall I compare thee to a summer's day?

William Shakespeare (Sonnet 18)

Summary

Chapter 10 covers making comparisons using graphics, complementing more formal statistical methods of comparison.

10.1 Introduction

Many scientific studies are carried out in a two-group design using a treated group and a control group. The groups are then commonly compared on the basis of the difference between their two means, using a two-sample t-test, assuming that the samples are drawn from normal distributions with equal variances. There may be other interesting features in the data, which cannot be formally compared because the sample sizes are too small. This should not stop us from looking at the data, and perhaps plotting the data should be compulsory if a t-test is carried out, just to remind people of the assumptions they are implicitly making and to make them think about what features they might be missing.

Many different features might be seen by looking at the data underlying t-tests. There may be evidence of differing variability, there may be outliers, either in relation to the whole dataset or in relation to one of the groups, there may be clustering in the data, there may be skewness, there might be favoured values. A t-test tells you nothing about any of these features and when the sample sizes are small, the situation the test was originally developed for by Gosset, there is too little data to be sure whether such features exist. As sample sizes get larger, it makes sense to look at more than just the difference between the means. Graphics and statistics are complementary approaches.

197

The dataset *bank* in the package **gclus** concerning forged and genuine Swiss banknotes was introduced in §5.6. Carrying out *t*-tests of the differences in means for the various banknote measurements comparing the genuine and forged notes gives a range of significances. Tests on the variables `Diagonal` and `Right` both give *p*-values of $< 2.2 \times 10^{-16}$. (This *p*-value limit may vary from machine to machine.)

The left-hand side of Figure 10.1 shows histograms of the `Diagonal` values for the two groups of notes and the right hand side shows histograms for the variable `Right`. The groups barely overlap for the `Diagonal` measurements, although there is one outlier amongst the genuine notes. On the other hand, the distributions of the `Right` measurements overlap a lot and it is mildly surprising that the means are so significantly different.

```
data(bank, package="gclus")
bank <- within(bank,
                st <- ifelse(Status==0, "genuine", "forgery"))
c1 <- ggplot(bank, aes(x=Diagonal)) +
             geom_histogram(binwidth=0.2) + facet_grid(st~.)
c2 <- ggplot(bank, aes(x=Right)) +
             geom_histogram(binwidth=0.1) + facet_grid(st~.)
grid.arrange(c1, c2, ncol=2)
```

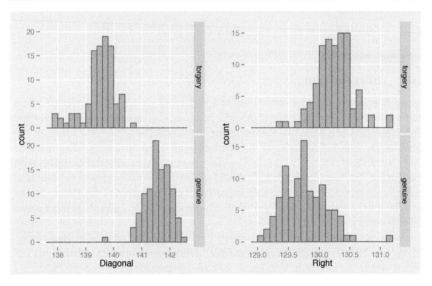

FIGURE 10.1: Histograms of the `Diagonal` measurements on the left and the `Right` measurements on the right for the forgery and genuine groups in the Swiss banknote dataset. *t*-tests confirm that the differences in means between the two groups are highly significant. The graphics show that the distribution patterns differ.

10.2   Making comparisons

"At the heart of quantitative reasoning is a single question: *Compared to what?*" [Tufte, 1990]. Numbers are of little use on their own, they have to be compared. In practice you are always making comparisons when you look at graphics, sometimes with explicit formal hypotheses, sometimes with implicit expectations.

If you are surprised by what you see, it is doubtless because you had anticipated seeing a different picture. For instance, a scatterplot with a completely random pattern can be very informative, if you believed that the data would be strongly correlated. An absence of information can also be information. In Conan Doyle's story "Silver Blaze", it was the fact that the dog did not bark in the night that Sherlock Holmes found curious. It is a good idea to think about what a graphic might look like before you draw it. We are all experts at explaining what we see—once we have the display in front of us.

Sometimes you compare new data to accepted standards, sometimes you compare new data with old data, mostly you compare data from different groups within the same dataset. Which comparisons you make, and why, can be crucial to the success of an analysis and there are often very many comparisons that might be considered. In a study of fuel use of cars you could compare this year with last year or earlier years, you could compare use in town traffic with motorway use, you could compare cars of different sizes and of different manufacturers.

There is also the issue of how you make the comparisons. Do you compare values or, if suitably paired data are available, study differences? And if you study differences, do you consider absolute differences or relative differences?

Groups may also be compared with descriptive statistics like medians, maxima, other quantiles, ranked values, or measures of variability. Medians might be used in comparing incomes, minima in comparing lap times, quantiles or ranked values in comparing top groups, and so on. Often the aim is to detect differences between groups, although sometimes it is to see if groups can be taken to be sufficiently similar to one another. For instance, an analysis of variance requires that the variability in each group should be approximately the same.

Of course, graphics are only part of the story; statistical comparisons have to be made. Usually means are compared, and the central limit theorem is often used, rightly or wrongly, as an all-embracing justification for that kind of testing. Alternatively non-parametric tests may be a possibility. Graphically there is more flexibility and it is possible to compare whole distributions, although there is less objectivity. One person's "it is obvious from the graphic that..." may be another person's "that could have arisen by chance and should be checked".

Types of comparison

With categorical data such as the Titanic dataset, discussed in Chapter 7, the main statistics for comparisons are rates and percentages. You might want to compare:

1. male and female survival rates;

2. the survival rate of the passengers compared to that of the crew;

3. survival rates between the three passenger classes;

4. survival rates by sex within class or by class within sex;

5. the survival rates with those of other sinkings (there has been research comparing the Titanic and Lusitania disasters [Frey et al., 2011]).

With all these possibilities it is important to choose appropriate comparisons. Given the survival rate of male adults in the third class, you could compare it with the rate for female adults in the third class or with the rate for male adults in the other classes. If there had been more male children on board you could further compare the male adult survival rate in the third class with the male child survival rate in the third class.

Comparisons may be

specific such as the male adult survival rate in the third class compared to the female adult survival rate in the third class;

general such as comparing the male adult survival rate in the third class to the overall survival rate (which includes that subgroup) or to the survival rate for all others;

at different levels comparing males and females, comparing male adults and female adults, comparing male adults and female adults in the third class.

Each comparison requires a different model with different standard errors and the interpretation of the results can be tricky given the large number of alternatives available for what you can compare.

In this chapter the emphasis is on using graphics to make comparisons. With formal statistical comparisons you have an explicit statement of what is to be compared. With graphical comparisons you decide what is important depending on what you see and on what your expectations were. In effect you may be unconsciously making many implicit comparisons, deciding what is worth looking at in more detail and what not. People may have different opinions of what they can see in a plot, although hopefully they will agree on the major features.

Comparing like with like

Comparisons should be fair, and that requires paying careful attention to several issues. Any differences identified should be due to the factor you want to investigate.

Comparable populations Samples provide information on the populations they are drawn from. Mortality and morbidity statistics are often reported as 'age-adjusted' to try to make them comparable across populations with different age structures.

Comparable variables In surveys, questions on the same topic may be framed in quite different ways. In business, markets may be defined differently by different companies, so that their sales figures are difficult to compare.

Comparable sources Data from different sources may be collected in different ways according to different rules. Opinion polls from different firms can differ systematically. Drink-driving laws and how they are implemented may vary widely between countries.

Comparable groups The gold standard in clinical research is the randomised control trial. Participants are allocated to treatments at random and, where possible, neither they nor the researchers know which treatment they receive (double-blind). This should go some way to ensuring that the groups are similar in all but the treatment they receive.

Comparable conditions If the effects of a particular factor are to be compared, then we need to ensure that the effects of all other possible influences are as far as possible eliminated. In the traditional discussions of the Hawthorne experiment it is concluded that the improvements in performance were probably due to the participants' being observed, not a factor the experimenters had originally taken into account.

Comparable measurements School marks in the United Kingdom are usually on a scale of 0 to 100. In Germany schoolchildren get grades from 6 (worst) to 1.

Standardising data Sometimes data are standardised to allow comparisons. In the decathlon event athletes' performances in each of ten disciplines are converted to point scores according to internationally agreed formulae, so that high jump performances can be compared with shotput performances, with sprint times and so on. Financial data collected over time are usually adjusted for inflation.

10.3   Making visual comparisons

Comparing to a standard

Michelson's measurements of the speed of light (used in Exercise 4 in Chapter 1) are shown in Figure 10.2 with a red dotted line added, showing the current estimate of the speed of light adjusted for travelling through air. Although the range of values obtained by Michelson contains the 'true' value, a 95% confidence interval based on these data does not. Using

```
data(Michelson, package="HistData")
tc <- t.test(Michelson, mu=734.5)
```

gives a confidence interval of 836.72 to 868.08.

```
ggplot(Michelson, aes(x=velocity)) + geom_bar(binwidth = 25) +
       geom_vline(xintercept = 734.5, colour="red",
       linetype = "longdash") +
       xlab("Speed of light in kms/sec (less 299,000)")
```

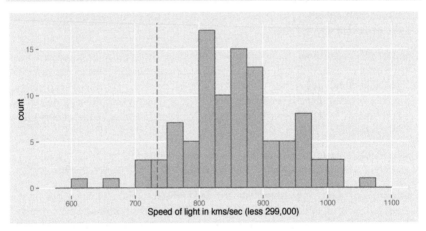

FIGURE 10.2: Michelson's data for the speed of light from 1879. Most of his observations were higher than the relevant currently accepted value, which would be 734.5 on this scale (marked in red).

Comparing new data with old data

There are at least two datasets with petrol consumption data for cars in R, *mtcars* with information on 32 models from 1973-4 and *Cars93* from the **MASS** package with information on 93 car models on sale in the USA in 1993 (also used in §5.4). Figure 10.3 compares a histogram of the miles per gallon reported for the cars in the earlier dataset with a histogram for the miles per gallon in city driving reported for the cars in the later dataset. The horizontal scales have been chosen to be comparable and the histograms have been drawn one above the other rather than side by side for easier comparison of the distributions. Even though city driving is more demanding than driving overall, the cars in the later dataset perform a little better.

```
data(Cars93, package="MASS")
c1 <- ggplot(mtcars, aes(mpg)) + geom_bar(fill="blue") +
            xlim(10,50) + xlab("mpg for 32 cars from 1973-4")
c2 <- ggplot(Cars93, aes(MPG.city)) +
            geom_bar(fill="red") + xlim(10,50) +
            xlab("mpg in city driving for 93 cars from 1993")
grid.arrange(c1, c2, nrow=2)
```

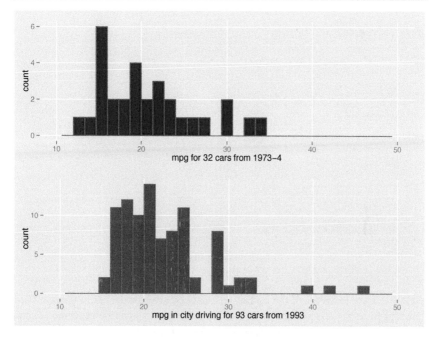

FIGURE 10.3: Comparing miles per gallon distributions for cars from 1973-4 and 1993. The later cars appear to have a slightly better performance and it should be taken into account that the mpg figures for them are only for city driving.

A comparison of the means with a *t*-test

```
tf <- t.test(Cars93$MPG.city, mtcars$mpg)
```

shows that the means are not significantly different, with a *p*-value of 0.067 and a 95% confidence interval of -0.16 to 4.71.

Comparing subgroups

Figure 10.1 showed comparisons of two of the variables from the Swiss Banknote dataset. Histograms are effective for comparing two groups and can reveal a lot of data details. They are less effective when there are several groups.

Boxplots work well for more groups. Figure 10.4 displays the `palmitic` variable for the nine areas in the *olives* dataset, which was also used in §9.3. Having the area labels directly beneath their corresponding boxplots helps a lot, as does colouring by region. For presentation purposes it could be worth ordering the areas by their medians or ordering them within the three regions by their medians.

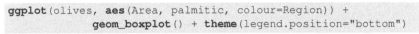

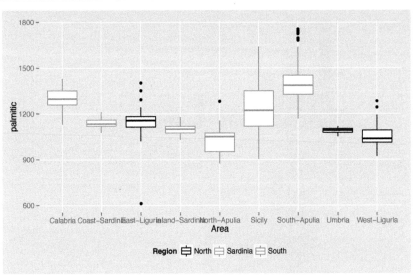

FIGURE 10.4: Boxplots of the `palmitic` data by area for the *olives* dataset. Differences in levels and variability as well as some outliers can be observed. There is much more variability in the Southern region than in the other two regions, both between and within areas.

Density estimates or possibly distribution functions can be used too, though for a smaller number of groups. Figure 10.5 shows boxplots, density estimates, and empirical distribution functions of the `palmitic` variable for the three regions in the *olives* dataset. Each display picks out the main feature (that the values for the Southern region are generally higher than those for the other two regions). The boxplots and density estimates are more informative than the empirical distribution functions.

If you amend the code above using `Area` instead of `Region` you can see how the comparisons of the nine areas would look using these plots. Although some features stand out, there are too many groups for an unambiguous allocation of colours, so that it is not always easy to tell which feature belongs to which area.

```
o1 <- ggplot(olives, aes(Region, palmitic, colour = Region)) +
          geom_boxplot() + theme(legend.position = "none")
o2 <- ggplot(olives, aes(x=palmitic, colour = Region)) +
          geom_density() + ylab("density") +
          theme(legend.position = "none")
o3 <- ggplot(olives, aes(x=palmitic, colour = Region)) +
          stat_ecdf() + ylab("cdf") +
          theme(legend.position = "bottom")
grid.arrange(o1, arrangeGrob(o2, o3, nrow=2),
          ncol=2, widths=c(1, 2))
```

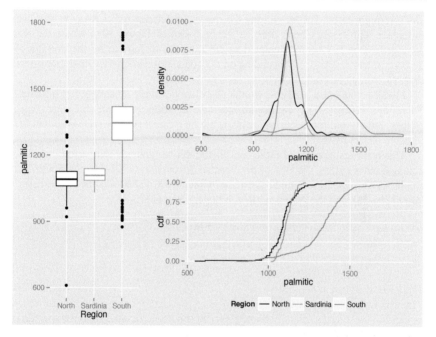

FIGURE 10.5: Three alternative displays of the distributions of the `palmitic` variable for the three regions in the *olives* dataset. The boxplots on the left emphasise the outliers and show that the olive oils from the South have higher values than the oils from the other two regions, while the values for the Sardinian region are very close together. The density estimates in the upper right plot show much the same information, although with less emphasis on the outliers, and reveal additionally that the distribution for the Southern region is skewed towards lower values. The empirical distribution functions in the lower right plot do not show much more than that the Southern region has higher values.

Comparing time series (Playfair's import/export data)

William Playfair drew many impressive displays, including several showing England's imports and exports with other countries. The data for England's trade with the East Indies between 1700 and 1780 have been estimated [Bissantz, 2009] from the graphic in the first edition of Playfair's *"Commercial and Political Atlas"*, published in 1786. The corresponding graphic in the third edition [Playfair, 2005] goes up to 1800.

There are several ways imports and exports could be compared over the years. Playfair plotted both series in the same display and coloured the area between them to give some idea of the balance of trade. The top plot of Figure 10.6 is a redrawing of his plot.

Cleveland suggested plotting the difference between imports and exports [Cleveland, 1994] and this is shown in the middle plot of Figure 10.6.

You might also consider plotting the relative difference rather than the absolute difference and this has been done in the final plot of Figure 10.6, where the relative difference has been calculated by dividing the difference by the average of `Imports` and `Exports`. Depending on the goals of the analysis, either `Imports` or `Exports` alone could have been taken as the base.

Modern economic analysts would probably consider adjusting for inflation. There are estimates of eighteenth century inflation available by year (e.g., [James, 2014]), but the cumulative effects are on a much smaller scale than the differences observed here.

This historical context of these data should not be ignored. The period includes, amongst other events, the War of the Spanish Succession (1701-14), the South Sea Bubble (1720), the Irish Famine due to the "Great Frost" (1740-41), the Seven Years' War (1756-63), and the American Revolutionary War (1775-83).

```
data(EastIndiesTrade,package="GDAdata")
c1 <- ggplot(EastIndiesTrade, aes(x=Year, y=Exports)) +
           ylim(0,2000) + geom_line(colour="red", size=2) +
           geom_line(aes(x=Year, y=Imports),
           colour="yellow", size=2) +
           geom_ribbon(aes(ymin=Exports, ymax=Imports),
           fill="pink",alpha=0.5) +
           ylab("Exports(red) and Imports(yellow)")
c2 <- ggplot(EastIndiesTrade, aes(x=Year,
           y=Exports-Imports)) + geom_line(colour="green")
c3 <- ggplot(EastIndiesTrade, aes(x=Year,
           y=(Exports-Imports)/((Exports + Imports)/2))) +
           geom_line(colour="blue")
grid.arrange(c1, c2, c3, nrow=3)
```

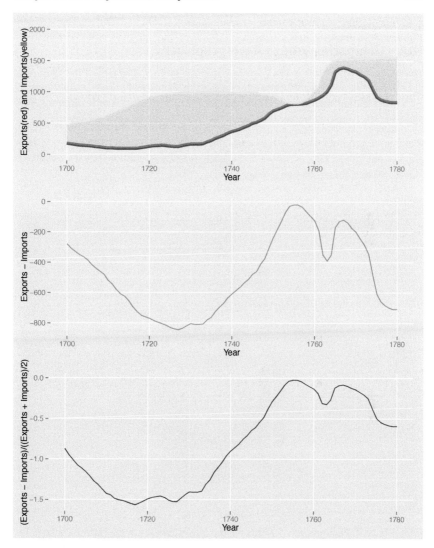

FIGURE 10.6: England's trade with the East Indies in the eighteenth century. The top plot is a redrawn version of Playfair's plot showing that imports were always higher than exports. The middle plot shows the balance of trade and highlights a dip in the 1760's, which is hard to see in the top plot. The lowest plot shows the relative balance of trade and suggests that recent deficits are lower, compared to the deficits in the twenty years around 1720.

10.4   Comparing group effects graphically

The famous *barley* dataset was considered briefly in §8.5. The aim of the study was to compare ten varieties of barley by looking at yields in two successive years at each of six testing station sites. The yields for the two years are a little different, with those for 1931 looking higher than those for 1932, as can be seen in Figure10.7.

```
data(barley, package="lattice")
ggplot(barley, aes(yield)) + geom_histogram(binwidth=5) +
    ylab("") + facet_wrap(~year, ncol=1)
```

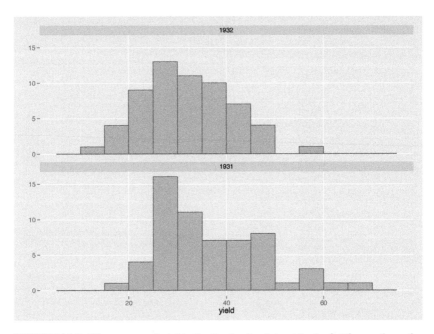

FIGURE 10.7: Histograms of yields for the *barley* dataset in the **lattice** package for the years 1931 and 1932. The values for the earlier year look higher on average with the distribution appearing to be shifted to the right compared to the following year.

It is the differences in yield by variety that are mainly of interest and drawing ten histograms would not be practical, even if there was enough data to justify it. Figure 10.8 shows parallel dotplots of the 12 values for the ten varieties in the upper plot and confidence intervals for the variety means in the lower plot. The plots suggest that there is little difference between the varieties. Note the use of the %>% operator from the **magrittr** package to build a sequence of operations in a readable order. **magrittr** is imported through loading the **dplyr** package.

```
c1 <- ggplot(barley, aes(x=variety, y=yield)) +
    geom_point() + ylim(10,70)
bar11 <- barley %>% group_by(variety) %>%
        summarise(N = n(), mean = mean(yield),
        sd = sd(yield), se = sd/sqrt(N))
lims <- aes(ymax = mean + 2*se, ymin=mean - 2*se)
p1 <- ggplot(bar11, aes(x=variety, y=mean)) +
    geom_point() + ylim(10,70) +
    geom_errorbar(lims, width=0.2)
grid.arrange(c1, p1)
```

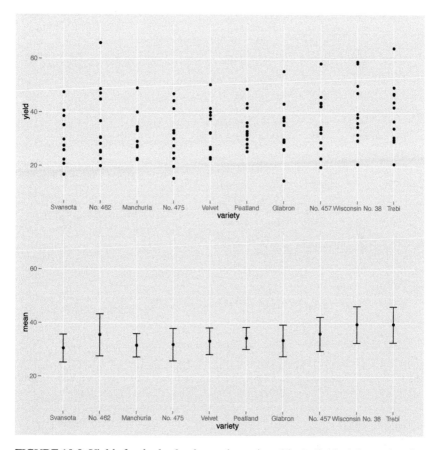

FIGURE 10.8: Yields for the *barley* dataset by variety. The individual data points for each variety are shown in the upper plot and confidence intervals for the means in the lower one (the intervals are $\pm 2SE$). The vertical axis scales have been chosen to include all the data and to be the same for both plots. The varieties appear to produce similar yields.

The displays in Figure 10.8 treat all values equivalently, ignoring both the sites and the years. Figure 10.9 shows confidence intervals for yields by year for each site. It is obvious that there are clear differences between the sites and between the years within the sites. The odd pattern for the Morris site that Cleveland remarked on, with the data possibly reversed for the two years, can also be seen.

The `mutate` function is used to put the two years in ascending chronological order, which looks more natural. It appears that the levels for the three variables, `variety`, `year`, and `site`, in the dataset in **lattice** have been ordered by increasing yields. You can confirm this by drawing boxplots of `yield` by each of the variables.

```
bar12 <- barley %>%
        mutate(Year = factor(year,
               levels = c("1931", "1932"))) %>%
        group_by(site, Year) %>%
        summarise(N = n(), mean = mean(yield),
           sd = sd(yield), se = sd/sqrt(N))
lims <- aes(ymax = mean + 2*se, ymin=mean - 2*se)
ggplot(bar12, aes(colour=Year, x=site, y=mean)) +
        geom_point() + geom_errorbar(lims, width=0.2) +
        ylim(10,70) + theme(legend.position = 'bottom')
```

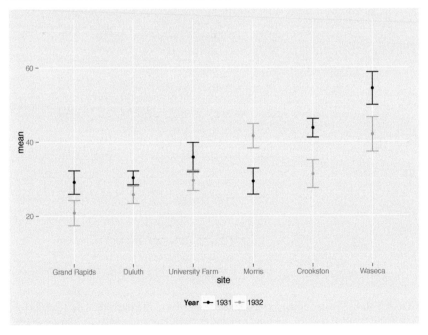

FIGURE 10.9: Confidence intervals for mean barley yields by year for each site. The data for 1931 are lower than for 1932 at all sites barring Morris.

To identify differences between the varieties, both site and year have to be taken into account to make an effective comparison. A linear model shows that variety is indeed significant and Figure 10.10 shows the interval estimates for the variety coefficients for this model. The default in R is to set the first coefficient to 0, so the plot shows that significance is primarily due to the last two varieties (Wisconsin No. 38 and Trebi) being higher than the first (Svansota).

```
m1 <- lm(yield~site+year+variety, data=barley)
library(coefplot)
coefplot(m1, predictors="variety", lwdOuter=1) + ggtitle("") +
        ylab("") + xlab("yield difference from Svansota")
```

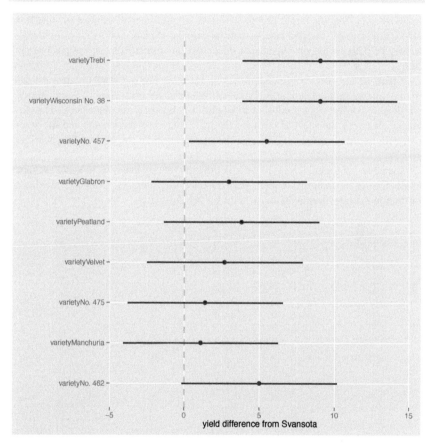

FIGURE 10.10: Interval estimates (±2s.e.) for the coefficients from a linear model of yield for the *barley* dataset. The top two varieties have coefficients that are obviously higher than 0, the value for the Svansota variety (not shown). The other varieties have positive coefficients, although not clearly higher than 0.

10.5   Comparing rates visually

As Chapter 7 shows, for example in Figure 7.1, shading bars of the same height
but different widths according to subgroup proportions is a good way of comparing
rates and the biggest groups get the most attention, being the widest bars. However,
there is no information on the possible significance of any differences between the
proportions. A statistical alternative would be to plot confidence intervals for the
proportions, treating them as estimates of unknown parameters.

Besides the advantage of offering some statistical guidance, there are two disadvantages, one major and one minor that have to be borne in mind. The minor
disadvantage is that the larger groups will have smaller intervals, thus attracting less
attention. The major disadvantage is that individual confidence intervals will not necessarily help assess the significance of the particular comparisons you might wish to
make.

Figure 10.11 shows confidence intervals for male survival rates by class for the
Titanic, assuming independent binomial distributions for each class. It is an alternative display for the right-hand plot in Figure 7.1, but only for the males. The differences between the survival rates for the male first-class passengers and the other
male groups are obviously significant; the significances of the other differences are
unclear.

The code for Figure 10.11 constructs the intervals by hand, so to speak. A new
dataset for just the males is defined. Then the necessary marginal totals are calculated
for finding the rates and from them the limits are calculated.

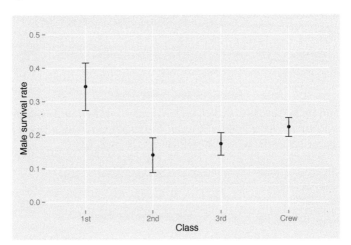

FIGURE 10.11: Approximate confidence intervals for male survival rates by class
for the *Titanic* dataset. The first-class males had significantly better survival chances
than the other males.

```
tl <- data.frame(Titanic)
tlm <- tl[tl$Sex=="Male",]
xt1 <- xtabs(Freq ~ Class, data=tlm[tlm$Survived=="Yes",])
xt2 <- xtabs(Freq ~ Class, data=tlm)
surv <- xt1/xt2
survS <- (surv*(1-surv)/xt2)^0.5
su <- data.frame(cbind(surv, survS))
su$Class <- rownames(su)
lims <- aes(ymax = surv + 2*survS, ymin=surv - 2*survS)
ggplot(su, aes(x=Class, y=surv)) + geom_point() +
       geom_errorbar(lims, width=0.1, colour="blue") +
       ylim(0,0.5) + ylab("Male survival rate")
```

Model-based confidence intervals for the rates could be calculated using logistic regression models of the males' data and the results would be roughly the same. The dataset would have to be rearranged first and after the model fitting the results converted to probability scales. Other logistic models could be fit as well, modelling the females as well as the males. A model with the interaction of Class and Sex then gives the same results, while a model with only the additive terms leads to different results. With displays of interval estimates you always have to know how the estimates were calculated.

```
library(reshape2)
tlc <- dcast(tlm, Class+Sex~Survived, sum)
mlc <- glm(cbind(Yes, No) ~ Class,
           family = binomial, data=tlc)
plc <- predict(mlc, se.fit=TRUE)
lowc <- plc$fit - 2*plc$se.fit
highc <- plc$fit + 2*plc$se.fit
estp <- 1/(1+exp(-plc$fit))
lowp <- 1/(1+exp(-lowc))
highp <- 1/(1+exp(-highc))
```

Alternatively, you could plot coefficient interval estimates from a model using **coefplot**, for instance with the following code. This uses a model form which includes no intercept, so that all the coefficients of interest are included.

```
tle <- dcast(tl, Class+Sex~Survived, sum)
mlf <- glm(cbind(Yes, No) ~ 0+Class*Sex,
           family = binomial, data=tle)
coefplot(mlf) + xlab("") + ggtitle("")
```

Rates can possibly depend on many, many factors. How baseball batting averages may depend on situational factors is discussed in [Albert, 1994], where eight factors including home/away, day/night game, groundball/flyball pitcher, scoring position/none are considered, and even more are suggested. Specific hypotheses require the appropriate underlying models for drawing intervals.

10.6   Graphics for comparing many subsets

Trellis displays (cf. §8.5) are an effective means of looking at displays of many subsets at the same time. Big differences for particular subsets stand out; smaller differences are more difficult to spot. Rearranging the order of the panels can help.
Figure 10.12 is a trellis display for the olive oil dataset showing scatterplots of the
variables `palmitic` and `palmitoleic` for the nine areas.

```
data(olives, package="extracat")
ggplot(olives, aes(palmitic, palmitoleic)) +
       geom_point() + facet_wrap(~Area)
```

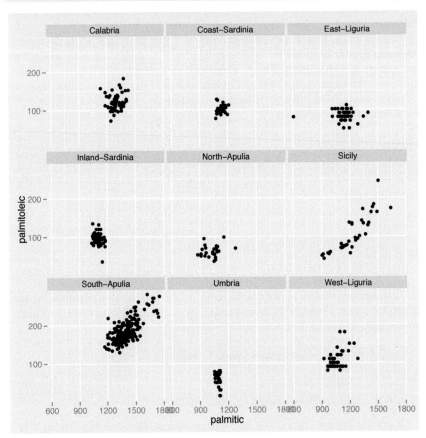

FIGURE 10.12: A trellis display of scatterplots for two of the variables in the *olives*
dataset. The data are in close clusters for several of the areas in different parts of the
plot. In two of the areas, Sicily and South Apulia, there are strong linear associations.

Trellis graphics can be drawn in R using the **lattice** package or the facetting options in **ggplot2**. It is interesting to compare the trellis graphic with a single scatterplot of the two variables in which the areas have been allocated different colours, Figure 10.13. Some features are easier to see, others not.

```
ggplot(olives, aes(palmitic, palmitoleic)) +
        geom_point(aes(colour=Area))   +
        theme(legend.position = "bottom")   +
        guides(col = guide_legend(nrow = 2))
```

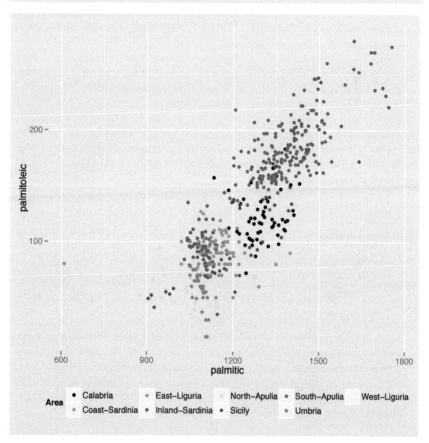

FIGURE 10.13: A single scatterplot for the data from Figure 10.12 with the points coloured by the variable `Area`. Overall there is a linear relationship between the variables `palmitic` and `palmitoleic` driven by the South Apulia region. Although the areas are separated to some extent, the density of points, overplotting, and the difficulty of distinguishing some of the colours make detailed interpretation tricky.

With interactive graphics you can select individual groups in a scatterplot and see them highlighted, while leaving the other cases in the background, providing context. A similar effect, but for all groups simultaneously, can be achieved by drawing a trellis plot with each panel displaying both its particular group highlighted in the foreground and also all the other cases in grey in the background. Figure 10.14 demonstrates the idea. It uses the function `facetshade` from the **extracat** package, which supplies the necessary additional facetting option for **ggplot2** objects. The `alpha` function from the **scales** package is used to set the alpha-blending of the points in the background.

It is now easier to judge the relative positions of the groups compared to the rest of the data than it was in Figure 10.12. A similar approach was used with some of the parallel coordinate plots in Chapter 6.

Figures 10.12, 10.13, and 10.14 offer three different ways of comparing scatterplots by groups. Personally I prefer Figure 10.14 to Figure 10.12, especially when `alpha` is chosen carefully, which may require a little experimentation. Figure 10.13 and Figure 10.14 are alternatives for different situations. The former is fine for well separated groups and when little display space is available (for whatever reason). The latter is better, when groups overlap. As always with facetting with no predefined order (the default is alphabetic), it is worth giving some thought as to whether another ordering might be more informative. For this dataset a grouping by the variable `Region` would make sense.

10.7   Graphics principles for comparisons

Graph size   Each graphic in a comparison must be drawn to the same size and aspect ratio. Comparing graphics of different sizes is possible, just more difficult.

Common scaling   The importance of using the same scales for groups being compared, common scaling, has been mentioned several times. It is easy to forget when using default options for displays. In general, both the horizontal and vertical axes should be the same for all graphics to be compared, although, as so often, the exception proves the rule. Histograms of two groups of different sizes may be better drawn with their own vertical axes if it is the form of the distribution that is to be compared rather than the data frequencies.

Alignment   Graphics to be displayed can be aligned vertically, which is good for comparing variability and form, or aligned horizontally, which is good for comparing the levels of peaks and troughs. Probably it is best to look at both. Either is much better than trying to compare without alignment.

Single and multiple windows   Plotting all groups in a single window can aid or hinder comparison. In a single graphic well-separated groups can easily be distinguished, whereas overlapping groups can be difficult to detect. Small multi-

```
data(olives, package="extracat")
library(scales)
fs1 <- facetshade(data = olives,
    aes(x = palmitic, y = palmitoleic), f = .~Area)
fs1 + geom_point(colour = alpha("black", 0.05)) +
    geom_point(data = olives, colour = "red") +
    facet_wrap(f=~Area, nrow=3) + theme(legend.position="none")
```

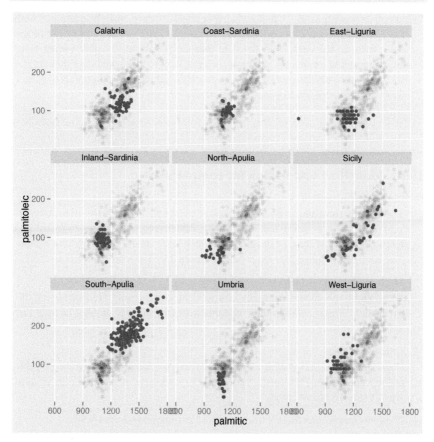

FIGURE 10.14: Another version of the trellis display in Figure 10.12. All cases are drawn in every panel with those for the corresponding subgroup drawn in colour on top and the remaining cases drawn in grey with alpha-blending underneath. This kind of plot shows the subgroups in context and makes it easier to assess the relative shapes of the subgroups. The West-Liguria oils have slightly higher palmitoleic values than the others and the East-Liguria outlier is an outlier for the whole dataset.

ples, with one display for each group, are the better alternative for overlapping groups. It can be helpful to draw the rest of the cases in the background to provide context as shown in Figure 10.14.

Colour and shape Distinguishing groups is helped considerably by using unique colours for each group. Shape and size are sometimes used for points, but colour is much more effective.

What is being compared A feature in a plot may subconsciously be compared with the rest of the data, with another part of the data, with several other parts of the data, with all the data, with the data that are graphically visible (when there are overlapping points, for instance), or even with remembered data that are not shown. If something interesting is found, it is valuable to determine which of these comparisons led to your thinking it was 'interesting'.

10.8   Modelling and testing for comparisons

1. Comparing means
   The default for comparing two means is the *t*-test. Even if the necessary assumptions are not ideally met, very low *p*-values are good evidence for real differences. You just have to decide what 'very low' means, although presumably everyone would agree that the *p*-values reported in §10.1 are.

2. More complex comparisons
   When additional factors are taken into account, comparisons are made using linear models. Testing should then take into account the large number of tests that are carried out.

3. Comparing rates
   Logistic regression is used for comparing two rates while including the possible influences of other explanatory variables. Proportional odds models may be used when there are ordinal rates.

4. Non-parametric approaches when standard assumptions do not hold
   There is a range of non-parametric tests which may be applied, including Wilcoxon for comparing two means and Kruskal-Wallis for non-parametric analysis of variance.

Multiple testing Looking for several kinds of difference at once amongst many different subgroups means that something of potential interest is bound to appear. This can even happen when looking at differences in rates between only a few groups: given m groups there are at least $\binom{m}{2}$ possible paired comparisons, and still more if comparisons of combinations of groups are included. The pragmatic approach is to examine the most extreme differences first and to bear in mind that the significance of any supporting statistical tests carried out will be affected by multiple testing.

There are formal methods for counteracting explicit multiple testing, such as Bonferroni corrections and FDR (false discovery rates). Details can be found in [Bretz et al., 2010] and [Benjamini, 2010]. These approaches do not cover graphical analyses. There are so many features that might be seen that you cannot be sure just how many tests are implicitly being carried out.

Main points

1. There is more to comparing groups than comparing their means. Quantiles, variability, and distributional patterns may also be compared—at least graphically (§10.1).

2. Fair comparisons need comparable data (populations, sources, variables, measurements, groups) and comparable graphics (size, aspect ratio, scaling, alignment) (§10.2).

3. Different comparisons show different aspects of datasets (Figure 10.6).

4. Comparisons may be made at different levels. Selecting the right comparison requires careful thought. Choosing appropriate conditioning variables is often important (Figures 10.8 and 10.9).

5. Appropriate intervals are needed for testing differences. Each comparison of interest may require its own model (§10.5).

6. Trellis scatterplots in which the rest of the data are plotted in grey in the background are effective for comparing overlapping subgroups (Figure 10.14).

Exercises

1. **Swiss banknotes**
   Consider the *bank* dataset from the **gclus** discussed in §10.1.

   (a) How do the distributions of the variable Length differ for the two groups defined by the Status variable? Draw histograms, boxplots, and empirical distribution functions. Which display do you find most informative? What are the advantages and disadvantages of the three displays in this application?

   (b) Are the mean lengths of the two groups significantly different?

2. **Petrol consumption**
   In §10.3, the petrol consumption of cars was compared for two datasets using miles per gallon figures. In most European countries, consumption is measured the other way round, as litres needed to drive 100 kilometres.

   (a) Draw comparative plots of petrol consumption, measured in gallons needed to drive 100 miles, for the two datasets. What features, if any, are notable in the plots?

   (b) Carry out a *t*-test comparing the two means. Discuss your result in conjunction with the result of the *t*-test carried out in §10.3.

   (c) A major influence on petrol consumption is the weight of a car. Draw scatterplots of MPG.city and 1/MPG.city against Weight for the *Cars93* dataset. What conclusion do you draw and which scatterplot do you prefer?

3. **Barley (corrected?)**
   Cleveland suggested that some of the data in the *barley* dataset analysed in §10.4 is possibly wrong. Construct a new version by switching the data for the Morris site for the two years.

   (a) Draw revised versions of the first two figures from the chapter. How different are your plots compared to the original ones?

   (b) Refit the linear model and plot the interval estimates. Do the conclusions about the differences between the varieties change with this revised dataset?

4. **Balance of trade**
   England's balance of trade with the East Indies was discussed in §10.3.

   (a) The relative balance of trade was calculated using the average of imports and exports. How would that graph look if you used (i) imports and (ii) exports as the denominator? How would you interpret the graphs and what headlines would you give them in a newspaper article?

   (b) What is the relation of imports to exports between individual countries nowadays? Find data for your own country and draw time series of the balance of trade between it and its three most important partners over the last twenty years. How have you defined "most important"?

5. **Diamonds**

The *diamonds* dataset from the **ggplot2** package includes information on over $50,000$ round cut diamonds.

(a) The variable `color` has categories ranging from D (best) to J (worst). Draw a plot showing how price varies with color. Are you surprised by the result? What might explain it?

(b) Draw a plot of the `color` coefficient estimates with 95% confidence intervals. What conclusions would you draw?

6. **Swiss banknotes (again)**

Consider the variables `Right` and `Left`, measurements of the edge widths of the notes.

(a) What do the distributions of the differences between these measurements for each note look like for the two groups? Are the differences significantly different from zero?

(b) The measurements `Bottom` and `Top` for the margin widths might also be expected to be close to equal for each note. Are they and does the difference relate to the edge width differences?

(c) Instead of using absolute differences, proportionate differences could be used. Draw a plot to compare the scales of the proportionate differences for the edges and margins. What denominator would you suggest? Do you think the data are reported precisely enough for these analyses?

7. **Olkin95**

There are data on 70 different studies of thrombolytic therapy after acute myocardial infarction in the *Olkin95* dataset in the **meta** package. (This dataset was also used in Exercise 2 in Chapter 5.)

(a) Plot the event rates for the experimental groups against the corresponding rates for the control groups. What does your plot show?

(b) The sizes of the studies should also be taken into account. Draw a scatterplot of the rate differences in each study against the size of the study, using the total number of participants for the size. (This is a kind of funnel plot.) What conclusions would you draw from your plot? How much does it matter, if at all, that the experimental and control groups are not always the same size?

8. **Intermission**

The *Judgement of Paris* by Rubens hangs in the *National Gallery* in London. How would you compare it with paintings of the same scene by, amongst others, Cézanne, Renoir, Raphael, and Watteau?

9. **Intermission (extended)**

Manet's painting *Le déjeuner sur l'herbe* is in the *Musée d'Orsay* in Paris. Monet painted a version a couple of years later and Picasso was inspired to make over one hundred drawings and some twenty-seven paintings of the scene. How do the versions by Picasso and Monet compare with Manet's?

11

Graphics for Time Series

The Government are very keen on amassing statistics—they collect them, add them, raise them to the nth power, take the cube root and prepare wonderful diagrams.

Josiah Stamp

Summary

Chapter 11 discusses drawing graphics for displaying the information in time series.

11.1   Introduction

Graphic displays are excellent for showing time series data and they are used extensively. Several of Playfair's graphics are of time series, and his displays of Balance of Trade data are well known [Playfair, 2005]. An example was shown in Figure 10.6. Economic and financial data are obvious applications for time series, but weather data, sports performances, and sales results all involve time series data too.

Graphics are particularly helpful for studying several time series at once. If each of a set of series can be transformed to a common scale, then they can all be plotted on the same display and it is possible to compare trends, changes in levels, and other features. If series show a common pattern for only part of the time period displayed, then that will be visible in the graphic while it may not be apparent in a statistical summary of the data.

Two entertaining and surprisingly informative time series graphics applications are Name Voyager [Wattenberg, 2005], which displays patterns of choice of baby names in the United States over the last 130 years, and Ngram Viewer [Google, 2010], which allows you to examine how word and phrase use developed over time in books. Much more can be found on visualising time in Graham Wills's book of the same name [Wills, 2012].

11.2 Graphics for a single time series

Consider the average number of goals scored per game in each of the first 46 seasons of the German Bundesliga from 1963/64 to 2008/09. In the first two seasons there were only 16 teams instead of 18, and in 1991/92 there were 20 teams. For that one year two extra teams were included to allow teams from the former East Germany to join. The averages can be calculated by aggregating the goals reported by team in the dataset *Bundesliga* in **vcd** and then dividing by the number of games. To display these data, there are several decisions to be made:

Symbol Should the data be represented by points and/or lines or by bars?

Scale What scale should be used for the y-axis, from what minimum level to what maximum level? Including zero on the y-axis, which is often rightly recommended in other circumstances, conveys the level of goals scored, but variability and possible trends may be shown less well.

Aspect ratio The look of a time series display is affected strongly by the aspect ratio, that is the length of the y-axis to the length of the horizontal time axis. The temptation to make a series appear flat (lengthen the time axis) or steep (shorten the time axis) should probably be resisted. [Cleveland and McGill, 1987] recommends choosing an aspect ratio that has the slopes of interest in the display at an angle of 45°. You just have to decide what the slopes of interest might be.

Trend Should a trend estimate in form of a smoother be added to the display? And, if so, which smoother should you use?

Gaps If the series is regular, how should gaps in the data be represented? Gaps could be time points when values were not recorded for some reason or they could be periods where there were no data to record, for instance holidays when shops are shut.

And there are additional, less critical, but still important decisions, such as the labelling of the time axis, the use of colour, and annotations referring to events at particular times.

Figure 11.1 shows the goals data using points and lines, a vertical scale slightly larger than the minimum and maximum, an aspect ratio with width twice height and including a spline smoother with pointwise 95% confidence intervals. The time axis is labelled by the first year of each season, i.e., the average goals per game for season 1970/71 are plotted at $t = 1970$. The plot shows that the average was up near 3.5 in the 1970s and early 1980s, then declined to under 3.0 and has been fairly constant since. There is no evidence that the introduction in 1995 of three points for a win instead of two made any difference to the numbers of goals scored.

Graphing the series for the average home goals and average away goals separately (Figure 11.2) reveals the interesting fact that the variation depended mainly on the numbers of home goals scored.

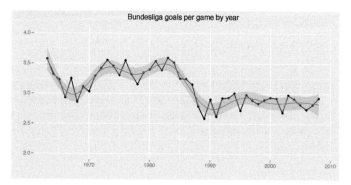

FIGURE 11.1: A time series of the average number of goals scored in the German Bundesliga over the first 46 seasons from 1963/64 to 2008/09 with a gam smoother added. There were 16 teams in the first two seasons and 20 in season 1991/92; otherwise there were always 18 teams. The average number of goals scored has been fairly stable since 1990, but at a lower level than in earlier years.

```
data(Bundesliga, package="vcd")
goals <- Bundesliga %>% group_by(Year) %>%
        summarise(hg=sum(HomeGoals), ag=sum(AwayGoals),
        ng=n(), avg=(hg+ag)/ng)
library(mgcv)
ggplot(goals, aes(Year, avg)) + geom_point() + geom_line() +
        stat_smooth(method = "gam", formula = y ~ s(x)) +
        ylim(2,4) + xlab("") + ylab("") +
        ggtitle("Bundesliga goals per game by year")
```

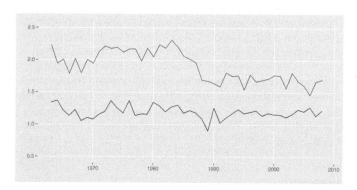

FIGURE 11.2: The average numbers of home (red) and away (blue) goals scored in the Bundesliga.

```
ggplot(goals) + geom_line(aes(Year, hg/ng), colour="red") +
        geom_line(aes(Year, ag/ng), colour="blue") +
        ylim(0.5, 2.5) + xlab("") + ylab("")
```

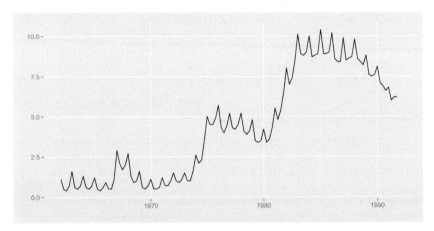

FIGURE 11.3: Unadjusted quarterly (West) German unemployment rates from 1961 to just after the unification of the two Germanies. Unemployment was very low in the 1960s and early 1970s, apart from a short spell at a higher level in 1967. There were distinct jumps in unemployment in the mid 1970s and in the early 1980s due to the oil crises of 1973 and 1979. Unemployment was declining at the end of the series from the high levels of the mid to late 1980s. The series shows a strong seasonal pattern with higher levels in winter than in summer.

```
library(zoo)
data(GermanUnemployment, package="AER")
ge <- as.zoo(GermanUnemployment)
autoplot(ge$unadjusted) + xlab("") + ylab("")
```

The *Bundesliga* dataset is an ordinary dataframe, but many time series datasets in R are provided in special time series classes, like the German quarterly unemployment data in the **AER** package. You can just use a default `plot` function or, alternatively, you can use the **zoo** package. It offers functions to convert other time series classes to its zoo class and an `autoplot` function to produce `ggplot` objects that can then be amended with other **ggplot2** options. This has been done in Figure 11.3 for (West) German unemployment data. The series shows a regular seasonal pattern and some relatively sharp changes in level between periods of fairly steady unemployment levels. The comparable unemployment rate in 2014, some 20 years later, is a little over 6%.

11.3   Multiple series

Various situations can arise when you have multiple time series, and they may need to be treated in different ways. You can have related series for the same population, such as death rates from different causes, and these can be plotted together or separately. You can have the same series for different subgroups, such as overall death rates for men and women, and they might be plotted together in a single display or individually in a trellis display. You can have series on quite different scales for the same population, for instance various economic indicators for a country, and these can be plotted in a single display with some kind of scale standardisation or each in their own window with their own individual scale.

Related series for the same population

The dataset *Nightingale* in **HistData** includes numbers of deaths in each month and monthly annualised hospital death rates for three different causes of death for the British Army in the Crimean War, from April 1854 to March 1856. The annualised rates have been calculated by multiplying the deaths in a month by 12 and dividing by the Army strength that month. Figure 11.4 shows the three rate series plotted together and it is obvious that deaths from disease dominate and that the winter of 1854/55 was a very bad time. That winter included the first half of the Siege of Sevastopol and the Charge of the Light Brigade at the end of October 1854. To put the numbers in context, 156 members of the Light Brigade died at the Charge [Wikipedia, 2014], while 503 members of the Army died in hospital from disease in that October—and that was one of the months with a relatively low death rate from disease!

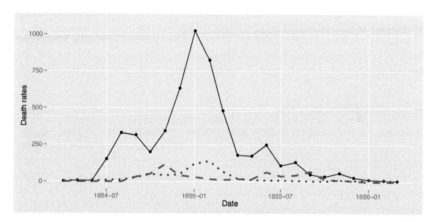

FIGURE 11.4: Annualised monthly hospital death rates per 1,000 in the British Army from disease (black line and points), wounds (red dashed line), and other causes (blue dotted line) in two years of the Crimean War, 1854-1856.

```
data(Nightingale, package="HistData")
ggplot(Nightingale, aes(Date)) +
        geom_line(aes(y=Disease.rate)) +
        geom_point(aes(y=Disease.rate), size=2) +
        geom_line(aes(y=Wounds.rate), size=1.3, col="red",
        linetype="dashed") +
        geom_line(aes(y=Other.rate), size=1.3, col="blue",
        linetype="dotted") + ylab("Death rates")
```

Instead of drawing each time series individually, there is an alternative approach, which can be more efficient. You first construct a new 'long' dataset of three variables: the time values needed, the values of the different time series all in one variable, and a grouping variable of the time series labels. The following code would produce a similar plot to Figure 11.4. The first line 'melts' the new dataset and the second part uses the grouping variable to draw the time series. There is no particular saving for this plot, but there would be if you wanted to display more series together.

```
library(reshape2)
Night2 <- melt(Nightingale, id.vars="Date",
   measure.vars=c("Disease.rate", "Wounds.rate", "Other.rate"),
   variable.name="NightV", value.name="Nightx")
ggplot(Night2, aes(Date, Nightx,
       colour=NightV, group=NightV)) + geom_line() + ylab("")
```

Same series for different subgroups

You can have time series of the same variable for several countries, giving time series that can be analysed together on the same scale, for instance GDP. Then you would have to decide if it is the total GDP's that are of interest (so the figures for the United States swamp those of many smaller countries), or GDP related to population size. An example was given in Figure 6.8 for the corn yields in the US by state over time. It is obviously sensible to study the bigger states first, but patterns over time for smaller states cannot be seen. More than one plot is needed, possibly an additional one for only the smaller states or one for each state separately using individual scaling.

Series with different scales

There may be several different time series for the same country: GDP and, say, imports, exports, and unemployment levels. Comparing patterns of movement in these series on the same graphic means finding some way to plot different variables on a common scale. One year can be chosen as a baseline and given a value of 100 with other years' values being transformed accordingly. That is often done with share prices and other financial variables. Alternatively, all series can be standardised by their respective means and standard deviations. This has the advantage of keeping series' levels and variabilities comparable but the disadvantage of changing with every additional time point.

Figure 11.5 shows an example of starting series off at the common value of 100. The data are the daily closing share prices in 2013 for four computing companies: Apple, Google, IBM, and Microsoft. The data were downloaded from Google Finance using the package **FinCal**. For this plot the vertical scale limits have been determined by default, using the overall minimum and maximum of the four series. Scale limits and the aspect ratio influence the look of time series a lot and it is usually worth experimenting with other alternatives. Apple started the year badly, but recovered to its original value. Google and Microsoft did very well. IBM started all right and then fell away and ended up about 10% down. Note the two sharp falls (Apple in January, Microsoft in the summer) and one steep rise (Google in the autumn).

A time series may have to be transformed to make it comparable with other series, as here, or even to make it more comparable with itself. Many monetary time series, such as salary levels or indeed GDP, are usually adjusted to take account of inflation.

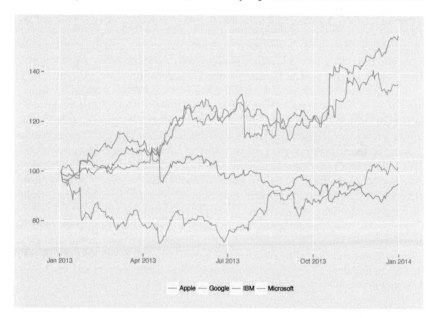

FIGURE 11.5: Closing share prices for four US computing companies in 2013. Each share price was set to 100 at the start of the year and the other prices transformed accordingly. Investing at the beginning of 2013, it would have been best to buy Google and Microsoft. Different conclusions would be drawn with other starting points.

```
library(FinCal); library(reshape2)
sh13 <- get.ohlcs.google(symbols=c("AAPL","GOOG","IBM","MSFT"),
                         start="2013-01-01",end="2013-12-31")
SH13 <- with(sh13, data.frame(date = as.Date(AAPL$date),
                              Apple = AAPL$close,
                              Google = GOOG$close,
                              IBM = IBM$close,
                              Microsoft = MSFT$close))
SH13a <- lapply(select(SH13, -date), function(x) 100*x/x[1])
SH13a <- cbind(date = SH13$date, as.data.frame(SH13a))
SH13am <- melt(SH13a, id="date", variable.name="share",
               value.name="price")
ggplot(SH13am, aes(date, y=price, colour=share,
       group=share)) + geom_line() + xlab("") + ylab("") +
       theme(legend.position="bottom") +
       theme(legend.title=element_blank())
```

One plot versus many

Displaying multiple series on the same plot works well if there are few crossings and if the individual series have low variability. Figure 11.4 is effective for showing the high death rates due to disease, but it would be less good for comparing the two series of death rates due to wounds and other causes. The four share price series in Figure 11.5 can be kept separate fairly easily, especially thanks to the use of colour. Figure 6.8 is also good at providing an overall picture, but does not represent series for individual states well. When there are many series, then a trellis plot could be used, with one panel for each series, possibly with the other series displayed with alpha-blending in the background, as was used in Figure 10.14.

The collection of datasets *hydroSIMN* in the package **nsRFA** includes the annual flows of 47 hydrometric stations in Piemonte and Valle d'Aosta. The data cover 65 years from 1921 to 1985, although only one series is complete and some are recorded for just a few years. Figure 11.6 shows all the series in one display. There appear to be some common features, such as the peaks in 1960 and 1962, and a few series seem to be partially grouped by colour. The hydrometric station code was made a factor so that more than one colour would be used, and the colours were assigned sequentially according to the numeric code. Stations with similar codes are generally close together spatially, which could explain the possible patterns.

Figure 11.7 displays the individual series. It is now much easier to see particular patterns and when data are available for each station. As well as indicating the differing levels and variability of series, the individual plots reveal the patterns over time better and suggest some possible outliers, for instance the first and last values for station 3.

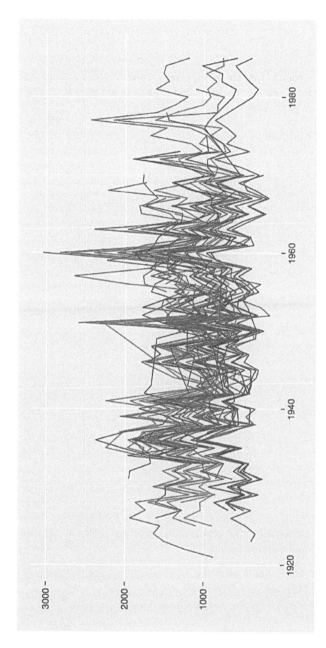

FIGURE 11.6: Annual flows for 47 hydrometric stations in Piemonte and Valle d'Aosta in Italy. There are common peaks, and stations that are spatially close are likely to have similar patterns. An accompanying map of where the stations are located would be very helpful.

```
data(hydroSIMN, package="nsRFA")
annualflows <- within(annualflows, cod <- factor(cod))
ggplot(annualflows, aes(anno, dato, group=cod, colour=cod)) +
    geom_line() + xlab("") + ylab("") + theme(legend.position="none")
```

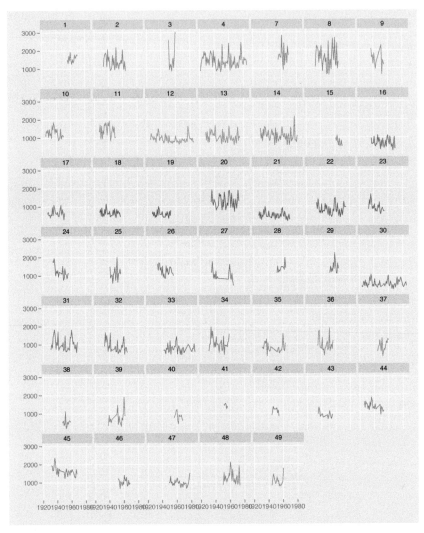

FIGURE 11.7: Annual flows for 47 hydrometric stations in Piemonte and Valle d'Aosta in Italy drawn individually. The changes over time are easier to see than in Figure 11.6 and individual patterns are clearer.

```
ggplot(annualflows,
       aes(anno, dato, group=cod, colour=cod)) +
       geom_line() + xlab("")  + ylab("") +
       facet_wrap(~cod) + theme(legend.position="none")
```

11.4 Special features of time series

Data definitions

Time series datasets differ from other datasets because of their dependence on time. The data have a given order and individual values are not independent of one another. In the simplest form there is a sequence of measurements of a single variable equally spaced in time, for example the annual GDP of a country. Even for this apparently straightforward situation it may be misleading to say "simplest". The definition of GDP might have changed over time, the value of money certainly has due to inflation, and the country itself might have changed. German data before and after 1990 can only be compared with caution, due to the reunification of the former East and West Germany.

In standard statistical analyses it is reasonable to assume that all data are defined similarly and are therefore directly comparable. That is not something that can be automatically assumed for time series. Unemployment statistics provide an interesting example, as the definition of who is unemployed varies across countries and is changed quite frequently within countries. The definition of unemployment in the United Kingdom changed over twenty times in the period from 1979 to 1993 when Margaret Thatcher was prime minister ([Gregg, 1994]). The change almost always resulted in a decline in the number of unemployed. The one change in definition in the other direction had actually been decided by the previous Labour government and only came into operation during Thatcher's premiership.

Length of time series

Time series can be short, for example, the annual sales of a new smart phone model, or long, for example the temperature values recorded every minute at a weather station over many years. Sometimes the short-term details of long series can obscure long-term trends, sometimes they are of particular interest. Plotting series on different time scales can be informative and for some financial time series such as exchange rates you could plot each minute, each day, each month, or each year. When you plot values for longer periods, there is a choice of what to plot for each period. You could take the final value, a value from the middle of the period, an average value, perhaps even a weighted average of some kind. Alternatively you could plot some kind of smooth. There are many possibilities.

Regular and irregular time series

In §6.5 parallel coordinate plots were used to plot regular time series, i.e., time series which are recorded at equally spaced time points. Hourly data, daily data, yearly data are often of this kind, but there are many series which are not. A history of a patient's temperature or blood pressure will rarely be based on a sequence of equally spaced time points. Political opinion polls are more frequent near elections than at other times. Sometimes treating the time points as equally spaced can be acceptable, often definitely not.

Plotting the time scale of an irregular time series accurately is important, and this is a situation where packages which handle times and dates correctly are a valuable help, especially as they allow you to plot several series with quite different time points on the same display. For example, a doctor might want to compare series of readings for different patients over a year, although he saw them at different times and different numbers of occasions.

With irregularly spaced series you cannot really speak of missing values and it is usual to just join successive time values. With regular series, it is a different matter. Missing values can then be awkward: should you leave a gap or simply join the points before and after with a straight line?

A related issue is time units that are thought of as the same, but are actually different, like months. February is always shorter than all other months of the year. Even the same months in successive years can be of different lengths. Retail sales could be some 3% higher for this month compared to the same month last year, because there was one extra shopping day.

Time series of different kinds of variables

Most time series are assumed to be of continuous variables, although you can also have time series of nominal variables (for instance, a person's state of health or each year's most popular fashion colour) or of discrete variables.

One of the most famous datasets in statistics is von Bortkiewicz's dataset on deaths from horsekicks in corps of the Prussian army. It is usually used to illustrate the Poisson distribution, as in §4.5, but the numbers of deaths are given over 20 consecutive years, so the data can also be viewed as a set of 14 time series, one for each corps. Figure 11.8 shows the total deaths each year. A barchart has been used to reflect the fact that the data are sums, single values for each year. The initial rise in deaths could be due to chance, better reporting, or larger corps sizes. The relevant data on corps sizes does not seem to be reported anywhere, despite the dataset being used in so many statistical textbooks.

```
data(VonBort, package="vcd")
horses <- VonBort %>% group_by(year) %>%
          summarise(totalDeaths=sum(deaths))
ggplot(horses, aes(year, totalDeaths)) +
       geom_bar(stat="identity") + ylim(0,20)
```

Outliers

Outliers in time series can be different from outliers for other kinds of datasets. They are not necessarily extreme values for the whole series, just unusual in relation to the pattern round them. A high temperature value in winter can be in the middle of the distribution of values for the whole year and yet still look way out of line for that

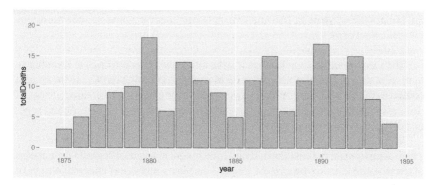

FIGURE 11.8: The number of soldiers killed by horse kicks in 14 corps of the Prussian army over 20 years at the end of the nineteenth century.

time of year. The famous *lynx* dataset (Figure 11.9) offers an interesting example. (Functions from the **zoo** package have been used as the dataset is a time series object of class ts.)

Should the two lower values in 1914 and 1915 be regarded as outliers, where something different was happening? The fact that this coincides with World War I makes it tempting to look for some kind of connection. Unusual values like that are easy to spot, although others may prove more difficult. Distances of points from curves are judged by the shortest line which can be drawn from the point to the curve. For time series, you have to assess the vertical distance between the point and the curve and this is hard to do for steeply rising or falling curves.

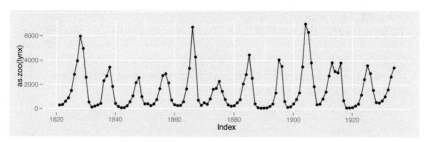

FIGURE 11.9: A time series of the number of lynx trapped on the Mackenzie River in Canada. There appear to be two kinds of cycles.

```
data(lynx)
library(zoo)
autoplot(as.zoo(lynx)) + geom_point()
```

Individual extreme values can influence scales and make the rest of a plot uninformative. This applies to time series just as much as to histograms, scatterplots, and parallel coordinate plots. However, scales for time series are unusual in that they are more likely to be influenced by values not included in the plot. If you display only the last couple of years of the exchange rate between the dollar and the euro, you may still want to let the vertical axis scale reflect the full history of the exchange rate since 2002.

The usual principle applies that it is best to draw several displays, zooming in to inspect details that may otherwise be hidden and zooming out to see the overall context.

Forecasting

There are two main reasons for studying time series: to try to understand the patterns of the past and to try to forecast the future. There are a number of alternative ways of displaying forecasts and it is good practice to make clear where the data end and the forecast begins, possibly using a change of background shading, a gap, or a dotted line into the future. Any forecast based on a model should include prediction intervals and including them in the display also emphasises where forecasts begin.

Figure 11.10 shows a two year forecast of the German unemployment data from Figure 11.3 using an exponential smoothing state space model. Detailed information and alternative forecasting models can be found in Rob Hyndman's **forecast** package.

As forecasts reach further into the future, the prediction intervals get wider and wider, which can make it tempting to use narrower ones. This temptation should be resisted. It is a sad truth that no matter how wide forecast intervals are, they may not be wide enough.

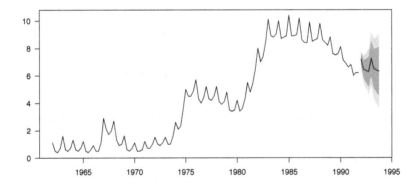

FIGURE 11.10: A forecast of (West) German unemployment for the years 1992 and 1993, based on the data from the previous 30 years. Prediction intervals of 80% and 90% are shown.

```
library(forecast)
par(las=1, mar=c(3.1, 4.1, 1.1, 2.1))
fit <- ets(ge$unadjusted)
plot(forecast(fit), main="")
```

Seeing patterns

It is remarkable how ready people are to see patterns in time series and to overlook features that are inconsistent with the supposed patterns. When series are plotted together, it is surprising, and even somewhat disturbing, how often people see causal relationships that may at best be associations due to other factors or just to the passage of time itself. And if you look at enough series, there are bound to be some that have something in common.

Yule gave an early warning example, the association between standardised mortality rates and the proportion of Church of England marriages over 46 years ([Yule, 1926]). Despite this, [Coen et al., 1969] reported amongst other results that they had found that the production of motor cars in the United Kingdom (seasonally adjusted) led the United States S&P stock market index by six quarters.

It would be tedious to list all the various issues to watch out for with time series, but it is well to remember that there are many.

11.5 Alternative graphics for time series

The best way to display time series is usually with lines joining the individual values. Only plotting the points can work as long as the variability around any trend is low. When a time series is short and what is being reported is a sequence of statistics rather than a continuous measurement, barcharts can be a good alternative. Figure 11.8 gives an example.

Parallel coordinate plots display regular time series well and can be readily used for plotting multiple series (§6.5). Their particular advantage lies in allowing you to inspect distributions of values across series at individual time points using boxplots.

When data are recorded daily and there are expected to be strong calendar effects, calendar plots can be a useful display. They are drawn just like a calendar, showing the twelve separate months and the weekly structure of the year. The rectangle for each day is coloured according to the value of the variable being displayed, so they are like a heatmap. Calendar plots pick out patterns in the weekly and monthly structure well, so they are good for checking the effects of weekends or special dates. As they use a colour scale, they work best when differences are large. The **openair** package offers a calendar plot and examples with **ggplot2** can be found on the web.

The association between two time series can be studied by plotting the two series together or by drawing a scatterplot, plotting one series against the other. That is what Yule did for the example mentioned in the last section [Yule, 1926].

11.6 R classes and packages for time series

Time and date variables require special treatment. Hours, minutes and seconds need
to be handled properly, as do months, changes from winter to summer time and back,
leap years, and time zones. R provides several packages which deal with these issues,
for instance **lubridate**, and the time series task view [Hyndman, 2013] is a good
reference. A number of packages offer special time series classes and the time series
task view is again the best source for information. There are tools to convert time
series objects from one class to another, as in the package **zoo**, to facilitate using
different classes and the **xts** package offers a single general class covering all.

Plotting tools in time series packages are particularly useful for labelling the
time axis appropriately, something which can be a difficult and frustrating task if
carried out from scratch. The website [TimelyPortfolio, 2013] gives a short historical
summary of time series plotting possibilities in R together with an example of the
same financial series plotted by the various packages discussed.

11.7 Modelling and testing time series

1. Single time series with regularly spaced time points
   Time series models generally assume equally spaced intervals between data
   points. Early on, attempts were made to decompose time series into four components: trend, seasonal, cyclical, and residual. Later on came the ARIMA models (autoregressive integrated moving average) promulgated by Box and Jenkins.
   With increasing interest in financial time series have come GARCH models (gen-
   eralised autoregressive conditional heteroscedasticity) and developments from
   them. And so it goes on, with everyone seeking the holy grail that will enable
   them to forecast the future successfully (and profitably).

2. Short, irregular time series
   Time series model fitting generally requires a lot of data. When series are both
   short and irregular, some kind of smoothing is a possible option. Figure 11.1
   shows an approach using spline smooths, which has the advantage of providing
   confidence intervals as well.

3. Multivariate time series
   Modelling of time series has concentrated on single time series. Modelling sev-
   eral time series together is a complex problem and progress has been slow.
   [Tsay, 2014] gives a thorough overview.

Main points

1. There are many different factors to bear in mind when drawing and interpreting time series (§11.2).

2. Single time series plots can show a lot of information (e.g., Figure 11.3).

3. Multiple series can be drawn in a single display to make comparisons easy (Figures 11.4 and 11.5).

4. Dates and times have tricky properties and it is best to take advantage of packages that can deal with them (§11.6).

5. Time series are a different kind of data and need to be treated specially. The data are not i.i.d. variables. Graphics are good for irregular time series and for displaying multiple time series. Models have difficulties with both. (§11.4)

Exercises

More detailed information for the datasets is available on their help pages in R.

1. **Air miles**
   The dataset *airmiles* is a time series of the miles flown annually by commercial airlines in the US from 1937 to 1960.

   (a) Before plotting the graph, think about what shape you would expect it to have. Plot the series and comment on the differences between what you get and your expectations.
   (b) Which aspect ratio conveys the information you find in the series best?
   (c) Do you think the graph looks better as a line graph (as suggested on the R help page for the dataset) or with points as well?
   (d) Might plotting a transformation help you to look more closely at the early years or would zooming in be sufficient?

2. **Beveridge Wheat Price Index**
   The Beveridge index of wheat prices covers almost four hundred years of European history from 1500 to 1869 and is available in the dataset *bev* in **tseries**.

   (a) Plot the series and explain why you have decided to plot it in that way.
   (b) Are there any particular features in the series which stand out? How would you summarise the information in the series in words?
   (c) Many important historical events took place over this time period, including the Thirty Years' War, the English Civil War, and the Napoleonic Wars. Is there any evidence of any of these having an effect on the index?

3. **Goals in soccer games**

   The *Bundesliga* dataset was used in §11.2.

   (a) Plot graphs of the rates of home and away goals per game over the seasons in the same plot. What limits do you recommend for the vertical scale?

   (b) Other possibilities for studying the home and away goal rates per game include plotting the differences or ratios over time and drawing a scatterplot of one rate against another. Is there any information in these graphics that is shown better by one than the others?

   (c) Can you find equivalent data for the top soccer league in your own country and are there similar patterns over the years?

4. **Male and female births**

   Important early demographic analyses were carried out on English data from the seventeenth century. The *Arbuthnot* dataset in the **HistData** package includes data on the numbers of male and female christenings in London from 1629 to 1710.

   (a) Plot the number of male christenings over time. Which features stand out?

   (b) Why do you think there was a low level of christenings from around the mid-1640's to 1660?

   (c) Two low outliers stand out, in 1666, presumably because of the Great Fire of London and the plague the previous year, and in 1704. A possible explanation for the 1704 outlier is given on the R help page for the dataset. Compare the data values for 1674 and 1704 to check the explanation.

5. **Goals in soccer games (again)**

   Consider the numbers of goals scored by each team.

   (a) How would you plot the annual average goals per home game for each team in the Bundesliga over the 46 seasons in the dataset? Would you choose a single graphic or a trellis display? Only one team has been a member of the Bundesliga ever since it started, Hamburg. How do you think the time series of teams with incomplete records should be displayed?

   (b) You could compare the annual home and away scoring rates of particular teams by plotting the two time series on the same display or by drawing a scatterplot of one variable against the other. Using the two teams Hamburg and Bayern Munich, comment on which display you think is better. Do the displays provide different kinds of information?

6. **Deaths by horsekick**

   Plot separate displays for each of the 14 corps in the von Bortkiewicz dataset (*VonBort* in **vcd**).

   (a) Do any of the patterns stand out as different?

   (b) 11 of the 14 corps had no deaths in the first year (1875). Could this be worth looking into?

7. **Economics data**

The package **ggplot2** includes a dataset of five US economic indicators recorded monthly over about 40 years, *economics*.

(a) If you plot all five series in one display, is it better to standardise them all at a common value initially or to align them at their means and divide by their standard deviations? What information is shown in the two displays?

(b) Alternatively you could plot each series separately with its own scale. Do these displays provide additional information and is there any information that was shown in the displays of all series together that is not so easy to see here?

8. **Australian rain**

The dataset *bomregions* in the **DAAG** package includes seven regional time series of annual rain in Australia and one time series averaged over the country.

(a) Can all seven regional series be plotted in one display or are individual displays more informative?

(b) Are there any outliers in the series and do they affect the scales used adversely?

(c) Is there any evidence of trend in the series? Are there cyclical effects?

9. **Tree rings**

The package **dplR** includes several tree ring datasets, including *ca533*. There are 34 series of measurements covering 1358 years in all from 626 to 1983. Note that no time variable is given, just the information that the data were recorded annually. The actual time range can be found from NOAA's tree ring database website.

(a) Plot all 34 series in separate displays. Are there any common features?

(b) There are at least two series with much higher maxima than the others. Compare a display excluding these series, but still retaining the same scaling for all the plots, with a display where each series is plotted with its own scale. What are the advantages and disadvantages of the two approaches?

10. **Intermission**

Salvador Dali's painting *The Persistence of Memory* is in the New York Museum of Modern Art. Do you think the distorted clocks could be interpreted as alternative models of time series?

12

Ensemble Graphics and Case Studies

Show and tell.

Common expression

Summary

Chapter 12 discusses using ensembles of graphics to explore datasets and contains a set of case studies for readers to investigate.

12.1   Introduction

When you first look at a dataset it is a good idea to draw many graphics to get a feel for the data. It is unnecessary to worry about labelling, legends, and the like, the aim is to quickly gain understanding of the information available. Resizing windows, varying aspect ratios, and redrawing displays with other formatting options all aid the process, and the windows can be casually spread around the screen or discarded. The many displays are for a single analyst's use, your own.

Saving graphics for future use or for presenting any features discovered to others is quite a different matter. Then you have to think about choosing effective versions of each graphic, of grouping them in a structured order, and of combining them in an ensemble, possibly linking them with text to tell a story. Carefully laid out combinations of graphics are primarily for presentation purposes. Tidying up graphics and planning a good layout can take up too much time during an exploratory analysis and, more importantly, a different kind of thinking is involved. Deciding on the technical details of drawing a display is not the same activity as studying the content of graphics displays to discover information. Drafting issues and computing concerns should not hamper graphical reasoning during exploratory analyses.

The examples in this chapter are all for presentation, intended to be readily understandable and to look reasonably presentable. A display of a whole raft of exploratory graphics would be complicated to explain and would represent only a momentary snapshot of an exploratory analysis. It is rarely crucial to describe how a result was obtained, but it is essential to be able to draw graphics which show that there is evidence to support the result.

Figure 12.1 shows three plots from the *coffee* dataset in the package **pgmm**. There are 43 coffees of two varieties, Arabica and Robusta, and the Robusta coffees are marked in red. The barchart shows how many coffees of each type there are, and that the great majority are Arabica. The parallel coordinate plot shows that there are several variables on which the two types differ and that the differentiation is particularly obvious on the variables Fat and Caffine [sic]. Finally, the scatterplot of those two variables confirms the main conclusion drawn from the pcp and presents it more definitively. In principle, all the information in the three graphics can be seen in the pcp. In practice, the barchart and scatterplot are valuable for emphasising the conclusions.

The code for Figure 12.1 is involved. Some of it is to specify the colours, some to reduce the amount of labelling, and some to draft the layout. The most important component of the pcp plot is the use of order to ensure that the lines for the group of interest are plotted on top of the lines for the rest of the data. Note that the spatial arrangement of the graphics and using a consistent style for them are just as important as the drawing of the individual graphics. For presentation you have to think of graphics as part of a story, not in isolation.

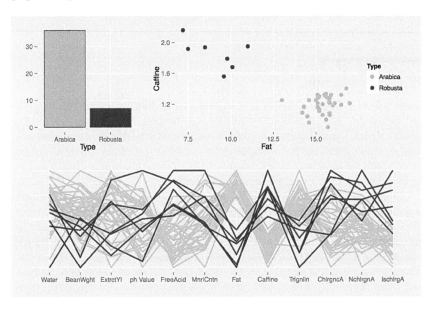

FIGURE 12.1: An ensemble of plots of the *coffee* dataset. Most coffees are of type Arabica (barchart). The Robusta variety coffees, coloured red in all three plots, are clearly different from the Arabica ones on two variables and fairly different on two others (pcp). The Robusta coffees have high levels of Caffine [sic] and low levels of Fat (scatterplot).

```
data(coffee, package="pgmm")
coffee <- within(coffee, Type <- ifelse(Variety==1,
                              "Arabica", "Robusta"))
names(coffee) <- abbreviate(names(coffee), 8)
a <- ggplot(coffee, aes(x=Type)) + geom_bar(aes(fill=Type)) +
          scale_fill_manual(values = c("grey70", "red")) +
          guides(fill=FALSE) + ylab("")
b <- ggplot(coffee, aes(x=Fat, y=Caffine, colour=Type)) +
          geom_point(size=3) +
          scale_colour_manual(values = c("grey70", "red"))
c <- ggparcoord(coffee[order(coffee$Type),], columns=3:14,
             groupColumn="Type", scale="uniminmax",
             mapping = aes(size = 1)) +
             xlab("") + ylab("") +
             theme(legend.position = "none") +
             scale_colour_manual(values = c("grey","red")) +
             theme(axis.ticks.y = element_blank(),
             axis.text.y = element_blank())
grid.arrange(arrangeGrob(a, b, ncol=2, widths=c(1,2)),
                         c, nrow=2)
```

To get a feeling for an exploratory analysis, think about how you might take a first look at the *coffee* dataset. Everyone will draw their own set of initial graphics, possibly including some of the following. You have to open a new graphics device for each new plot (e.g., with dev.new()) or earlier ones will just be overwritten. Not all of these graphics are nice to look at, although they can be improved by resizing the windows appropriately and you can then better gain information from them.

```
#Barcharts of the categorical variables
ggplot(coffee, aes(factor(Variety))) + geom_bar()
#Comment: More Arabica than Robusta
ggplot(coffee, aes(Country))  + geom_bar()
#Comment: Many different countries (with odd spellings)

#Checking for missings
visna(coffee)     #Comment: There are no missings

#Mosaicplot of the categorical variables
mosaic(Variety ~ Country, data=coffee)
#Comment: Only one country has both varieties

#Distributions of the continuous variables
boxplot(scale(coffee[,3:14]))
#Comment: A few outliers, mainly symmetric, try histograms?

#Scatterplot matrix of the continuous variables
pairs(coffee[,3:14], pch=16)
#Comment: A few interesting patterns, too many plots
```

12.2   What is an ensemble of graphics?

Many of the examples up till now have used single graphics, sometimes in conjunction with alternative graphics. It is overly ambitious to think you can discover or represent all the interesting information in a dataset with one graphic, and for real applications it is certain that you will have to draw a large number of them. Perhaps everyone does this in practice, even if it is rarely described or discussed. In the distant past this may have been due to the difficulty of drawing graphics, when that had to be done by hand, and in the more recent past to restrictions of space for printing graphics in published work. Neither of these constraints exists today. Graphics may be drawn quickly and easily, and printed publications may be supplemented by material provided on supporting webpages.

Ensemble modelling methods have become increasingly used in statistics and machine learning. Combining many models, sometimes quite simple ones, is often more effective than a single complex model. The same applies to graphics. Ensembles of graphics are groups of displays for presenting several different aspects of a dataset simultaneously. They should be viewed as a group, with each contributing something else to the overall picture. You might have

- Several plots of the same type for the same variable(s)
  It is often informative to draw several histograms with different binwidths, for example Figures 3.2 and 3.3. Groups, modes, and favoured values stand out with some scalings more than with others.

- Several different plots for the same variable(s)
  Boxplots, histograms and density estimates emphasise different features of the data, for example Figures 3.11 and 3.12.

- Several plots of the same type for different subgroups (small multiples)
  Trellis plots are the standard when the combined subgroups cover all the dataset, e.g., Figures 10.12 and 10.14. Sets of subgroup plots are good for making comparisons.

- Several plots of the same type for different though comparable variables
  If exam marks are available for a number of different subjects (as in the *mathmarks* dataset in **SMPracticals**), then a set of histograms, one for each exam, all scaled the same, give a good overview and enable comparisons across subjects. Plots of multiple time series also fall into this category when they are plotted individually (as in Figure 11.7).

- Several plots of the same type for different variables
  Figure 3.9 presents all the one-dimensional marginal distributions for the Boston housing dataset. The distributional shapes give an initial impression of the types of information the variables provide, some grouped, some with outliers, some skewed. Scatterplot matrices are another example and can be seen in Figures 1.9 and 5.13.

- A variety of plots for different variables
  Information derived from studying one set of variables may be checked by looking at other variables. Sometimes the same types of displays are useful, sometimes others are. The information you discover in different parts of the dataset should be put together to construct an overall picture. Examples include Figure 9.6 and Figure 12.1. Many scientific reports use ensembles of plots and you can find plenty of examples in issues of the weekly scientific journal Nature. For presentation purposes it can be helpful to combine plots to tell a story. Some of the best information visualisation presentations use this approach, as you can see in the work done by the New York Times graphics group [New York Times, 2011]. Some of these presentations are put together to form a narrative flow, an attractive way of presenting information in a convincing manner.

Combining graphics of different sizes and aspect ratios in one display needs planning. It is probably best to sketch what you want first and then choose suitable units to match the layout. Histograms tend to be wide and not high, scatterplots square, boxplots narrow and high, pcp's short and very wide, and the form of barcharts depends on the number of categories. R is flexible enough to allow any mixture, but the success of the design depends on you.

The term 'ensemble of graphics' implies a static collection of graphics put together to present a coherent story. In practice, with exploratory graphics, ensembles are more a matter of keeping several balls in the air at once, keeping track of a range of lines of thought in parallel as a graphical analysis proceeds. How is the feature apparent in one graphic expressed in another? Are the changes in one variable consistent with the changes in another? Is there an interesting feature, which needs to be checked with a new display? Which features are most striking and should be followed up? What additional graphics are needed to attempt to answer the new questions which have arisen? With the benefit of the knowledge you have discovered, it can be simple enough to put together the graphics that reveal the information clearly and convincingly, you just have to find the information first.

12.3   Combining different views—a case study example

The *Fertility* dataset in the **AER** package concerns a study of women with at least two children. It was a large study with some 250,000 women involved. The interest was in which mothers had more than two children. The explanatory variables provided include the genders of the first two children, the mother's age and race, and how many weeks she worked in the previous year (1979).

Figure 12.2 shows an ensemble of plots for the dataset. The code makes use of **ggplot2** objects to simplify drawing several plots of similar type and uses the `gridArrange` function from the **gridExtra** package to design the somewhat elaborate layout.

As expected, the numbers of women with two or more children rise with age, although it is perhaps surprising that the pattern is so regular before dropping off for the two oldest cohorts. The same pattern exists for all race groups (because of overlaps there are six), as can be seen using

```
data(Fertility, package="AER")
ggplot(Fertility, aes(x=age)) + geom_bar(binwidth=1) +
    facet_wrap(~ afam+hispanic+other, scales="free_y", ncol=8)
```

The function `facet_wrap` ignores any empty combinations while the option `scales="free_y"` scales the y axis of each graph individually, so that distributional shapes are compared rather than absolute frequencies.

That the majority of women either work or do not work is to be expected. Using a function like

```
with(Fertility, prop.table(table(work)))}
```

shows that just over a third reported something in-between.

From the barcharts in Figure 12.2 you can see that most of the women had just two children, that boys were as usual slightly more common than girls, and that most women in the study were Caucasians.

The numbers of women in the study increase with age and so does the proportion of them with more than two children. This can be seen in Figure 12.3. Interestingly, more women have three children if the first two have the same sex and this is irrespective of whether the first two were boys or girls (Figure 12.4).

The *Fertility* dataset is very large and so it is possible to look at the combination of the effects of age and the gender of the first two childen. Figure 12.5 shows that the pattern is pretty much the same for all ages.

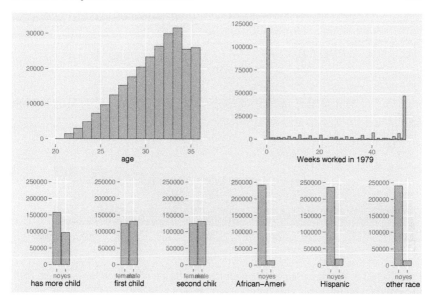

FIGURE 12.2: Plots of the variables in the *Fertility* dataset. The numbers of women with two or more children rose steadily with age, except for the eldest two years. The majority of women worked every week of the year or not at all. Fewer women had more than two children than had exactly two. There were slightly more boys than girls for both the first and second child. Most women in the study were Caucasian.

```
data(Fertility, package="AER")
p0 <- ggplot(Fertility) + geom_bar(binwidth=1) + ylab("")
p1 <- p0 + aes(x=age)
p2 <- p0 + aes(x=work) + xlab("Weeks worked in 1979")
k <- ggplot(Fertility) + geom_bar() + ylab("") + ylim(0,250000)
p3 <- k + aes(x=morekids) + xlab("has more children")
p4 <- k + aes(x=gender1) + xlab("first child")
p5 <- k + aes(x=gender2) + xlab("second child")
p6 <- k + aes(x=afam) + xlab("African-American")
p7 <- k + aes(x=hispanic) + xlab("Hispanic")
p8 <- k + aes(x=other) + xlab("other race")
grid.arrange(arrangeGrob(p1, p2, ncol=2, widths=c(3,3)),
            arrangeGrob(p3, p4, p5, p6, p7, p8, ncol=6),
            nrow=2, heights=c(1.25,1))
```

```
doubledecker(morekids ~ age, data = Fertility,
             gp = gpar(fill = c("grey90", "green")),
             spacing=spacing_equal(0))
```

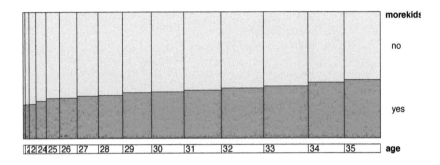

FIGURE 12.3: A spinogram of age with the proportion of women having more than two children highlighted. This proportion increases with age.

```
doubledecker(morekids ~ gender1 + gender2, data = Fertility,
             gp = gpar(fill = c("grey90", "green")))
```

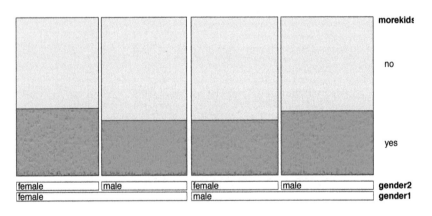

FIGURE 12.4: A doubledecker plot showing the numbers of women having two girls, girl boy, boy girl, or two boys for their first two children. The proportion of women having more than two children is highlighted. Mothers of same sex children are more likely to have more than two.

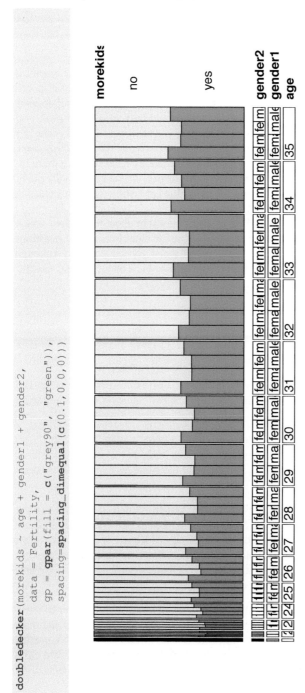

FIGURE 12.5: A doubledecker plot by gender pattern for the first two children and age. The proportion of women having more than two children is highlighted and you can see that the same pattern of higher rates for same sex families holds true for all ages.

12.4   Case studies

The following case studies are for you to investigate, so you can test your graphical skills. More information on the datasets involved can be found on the corresponding help pages in R. Full information on how and why the data were collected is sometimes provided. Ideally for any study there should be a clear set of goals and close cooperation with subject area experts. Even if you have well-defined goals, it is still sensible to do some exploring around. Sometimes issues of data quality become apparent, sometimes a question about an odd minor feature leads to much more information being provided about other parts of the dataset. As Tukey famously put it, data analysis is detective work, and we all know that apparently innocuous clues in detective stories can lead to major discoveries.

Each case study begins with a group of direct questions about the dataset, which should be relatively straightforward to answer. One or more open-ended questions follow that will hopefully encourage you to see what else you can discover. Although analyses should be carried out graphically, you can also think about how any conclusions could be checked with models.

Moral statistics of France

The *Guerry* dataset in the package **HistData** includes a range of information for each of the 86 departments of France around 1830.

1. There are two variables relating to donations, `Donations` (to the poor) and `Donation_clergy`. Draw histograms of the two variables and comment on the differences between them.

2. What would you expect the scatterplot of the last two variables, `Prostitutes`, numbers of prostitutes in Paris by department of birth, and `Distance`, distance from Paris, to look like? Does a scatterplot of the variables match your expectations?

3. Draw a boxplot of the variable `Infants`, which refers to the population per illegitimate birth. Which of the departments marked as outliers do you think are definitely outliers? If you invert the variable to get a measure of illegitimate births by population and draw a boxplot, other departments are identified as possible outliers. Which of these do you think are definitely outliers? Which variable would you judge to be more appropriate to use?

4. Some of the variables are expressed as inverses of rates, so that higher is better. Quite a few variables are only given as ranks, so that lower is better and there is only ordering information, but no distributional information. Draw a parallel coordinate plot of the rank variables and select the departments with big cities (category `3:Lg` in the variables `MainCity`). How do these departments differ from the others, if at all?

5. What additional features can you discover in the dataset?

Airbags and car accidents

There are 26217 cases of police-reported car crashes in the *nassCDS* dataset in the package **DAAG**.

1. Draw a histogram of the variable `weight`. The R help file for the dataset says that the observation weights are "of uncertain accuracy". Is there any evidence of this? What graphics would you draw to investigate which cases have high weights and which have very low weights?

2. How does the availability of airbags depend on the age of the vehicle?

3. How does death rate depend on vehicle speed?

4. How does death rate vary with the variables `seatbelt`, `airbag`, `deploy`, and `frontal`? Which orderings of the variables and of the categories within the variables give the most convincing graphic?

5. Are there other interesting patterns in the data worth presenting?

Athletes' blood measurements

The *ais* dataset in the package **DAAG** contains five blood measurements and six physical measurements for 202 male and female athletes.

1. Draw histograms of the eleven variables. Are there many different forms or are they all roughly normal?

2. Boxplots of the variables suggest that there are a number of outliers. How many of these are outliers on more than one variable?

3. Draw a parallel coordinate plot of the eleven variables with the females coloured red. Do the females have lower values than the males on all the variables? Which variables look most promising for distinguishing between males and females?

4. There are ten different sports listed, nine for the women and eight for the men. Is there anything particularly distinctive about the patterns for any of the sports, either for both genders together or singly?

5. If you were to choose three graphics to represent the most interesting information in this dataset, which would you choose and how would you present them in a summary report?

Marijuana arrests

The *Arrests* dataset in the package **effects** contains information on over 5000 arrests on the streets of Toronto over five years. There are six variables describing the arrested person, a variable giving the year, and a binary variable `released` saying whether the person was released with a summons or not.

1. Are there any features worth mentioning in the distributions of the individual variables? Which graphics show these features clearly?

2. How does the variable `released` depend on the others, taking them one at a time? Which graphics would you select to show this?

3. What about the dependence of `released` on the variables `colour`, `employed`, and `citizen` in combination? Which plot would you draw to investigate this?

4. Are there other features of the dataset worth reporting?

5. If you wanted to summarise the information you have found on one page, which graphics would you choose and how would you present them as a group?

Crohn's disease

Crohn's disease is a chronic inflammatory disease of the intestines. The package **robustbase** includes a small dataset *CrohnD* of 117 patients in three groups, one placebo group and two different drug groups.

1. Draw plots to show there are equal numbers in the three treatment groups, but not equal numbers by gender or country.

2. Which plot would you draw to show the counts for all gender, country, and treatment combinations?

3. The key outcome variable is the number of adverse events, `nrAdvE`. What graphic would you choose to display its distribution? What about a set of three graphics, comparing the distributions of the numbers of adverse events for the three treatment groups?

4. Draw histograms of the other five variables, including `ID`. Explain why you chose the binwidths you use and describe what information you can see in the displays.

5. If you had to write a short report on your analysis, what would you include? (There is some interesting background information in [Brant and Nguyen, 2008] and there are plenty of other sources of information you might find useful.)

Footballers in the four major European leagues

Statistics has even reached soccer nowadays and the dataset *EURO4PlayerSkillsSep11* in the package **SportsAnalytics** includes 43 pieces of information on 1851 soccer players in September 2011.

1. Draw a scatterplot of the variables `Attack` and `Defence`. Are there any obvious outliers? What might be done about them?

2. There is a variable called `Position` saying whether players are goalkeepers, defenders, midfielders, or forwards. Which group of players scores highest on the variable `Agression`?

3. Draw a graphic showing the relationship between the variables `Foot` and `Side`. What are your conclusions? Do your conclusions vary depending on a player's position?

4. How would you display graphically how goalkeepers differ from other players?

5. Consider only the subset of players who are goalkeepers. What sort of values are they given for variables such as the four concerned with passing? Do these distributions depend on the league in any way?

6. The dataset has many variables and so there is much that might be investigated. Choose one of the four leagues and investigate if there are any noteworthy differences between the teams. If teams are distinctive in some way, check whether they are also distinctive if you take the other leagues into account as well. (The data are provided to support a soccer video game and the way the players are evaluated is not clearly documented. This may account for some of the minor features you can find.)

Decathlon

Both from a sporting point of view and from a statistical point of view, the decathlon is a fascinating event. Athletes must compete in ten different disciplines over two days. Their performances in each event are transformed to a common scale and summed to produce a total score.

The transformation formulae [IAAF, 2001] were last changed in 1985 by a committee including the academic and former decathlete, Viktor Trkal. That their recommendations have continued to be used for so many years is a tribute to their work. Viktor Trkal [personal communication] reported that the committee used principles they believed a satisfactory system should follow and data provided by various countries on decathlon performances to guide their work. One principle that was not respected by the previous system was that an improvement in performance should be worth more points the higher the level that is improved on.

There is an excellent webpage for decathlon results, based in Estonia, where individual scores and much additional information can be found for performances over many years [Salmistu, 2013]. The dataset used here, the *Decathlon* dataset in the package **GDAdata**, is a subset of that data for the years 1985 to 2006 covering almost 8000 performances. The criteria for a score being recorded is that the total points should be higher than 6800 and that it should be the best performance of an athlete in a particular year. This means that the top decathletes can only appear at most once a year, although they can and do appear in several different years.

1. What plots would you choose to study the variable Totalpoints and how would you describe the distribution?

2. Use boxplots to examine how the distribution of Totalpoints develops over time. What conclusions would you draw?

3. Draw a scatterplot of the variables m100 and P100m, the actual times in seconds, and the points awarded for the 100 metres. How close is the formula to a linear function in the range covered by the data? Does the same apply to the pole vault event?

4. Draw parallel coordinate plots of the raw data (m100 to m1500) and the points data (P100m to P1500). Draw boxplots for the points variables. What information can you see in each plot? Which plot is most useful for comparing the performances in the different disciplines?

5. Pcp's of the points variables on a common scale could be sorted by the maximum values, the medians, the IQR's, or by those statistics for a selection, such as for the most recent year. Choose any two you find interesting and explain your choice.

6. For the second question you looked at the development of the total points scored over the 22 years. What about the development of scores for the individual disciplines? You might compare the maximum, the median, the 95% quantile, the tenth best performance or the twenty-fifth best performance for each year. Which do you think would be suitable? Create the necessary dataset to plot the statistics you chose over time. What conclusions do you draw from your plot?

Intermission

Rembrandt's *The Night Watch* hangs in the *Rijksmuseum* in Amsterdam. How might you divide it up into individual groups and scenes? How well do individual scenes combine to form the whole?

13

Some Notes on Graphics with R

Those are my principles, and if you don't like them... well, I have others.

Groucho Marx

Summary

Chapter 13 discusses using R for graphics and passes on some individual pieces of information that have been helpful for examples in the book.

13.1   Graphics systems in R

There are many different ways of achieving the same goals in R and you have to decide for yourself which works best for you. This may depend on flexibility, speed, elegance, or ease of use. The standard R graphics are called **base** graphics and were written to match the graphics available in S. They offer extensive control over the look of the displays. This makes them very flexible and suitable for presentation purposes. S graphics were advanced for their time, but computing has moved on considerably since then in both hardware and software terms. An alternative graphics system, grid graphics, implemented in the package **grid**, is described in [Murrell, 2011]. This provides a systematic underlying structure and increases user control further. Both of the popular packages **ggplot2** and **lattice** make use of **grid**.

   R is like a spoken language, in that there are a huge number of ideas you can express and there are multiple ways of doing that. In graphics terms this means that there are many different displays available and often several alternative ways of drawing them. For graphics examples you can peruse the R Graphical Manual (rgm3.lab.nig.ac.jp/RGM/), which collects R graphics that developers and users have drawn and the code written to produce them. As for all the different ways of drawing particular plots, just think of the various histogram and scatterplot matrix possibilities. There is a task view for graphics, [Lewin-Koh, 2013], which lists many packages and gives a helpful overview. In this book the package **ggplot2** has mostly been used.

13.2   Loading datasets and packages for graphical analysis

Some datasets are supplied in the package **datasets** and are directly available. Others are supplied as parts of packages and you have to have the package installed to get the dataset. The dataset can then usually be loaded using

```
data(datasetName, package = "package_name")
```

Naming the package ensures that you get the dataset you want. The same dataset can be contained in more than one package in different formats and yet the identical name. And in some cases quite different datasets have identical names in different packages (e.g., *movies* and *barley*). Note that it is not necessary to load a package with `library(package name)` to make a dataset in the package available. Loading a package unnecessarily can have the potentially irritating side-effect of having functions unintentionally (and confusingly) overwritten.

In a similar way, some functions in different packages have the same name, although they do quite different things. To avoid loading a complete package you can use `package::function`. This was done in the code for Figure 9.1, since the function was only needed for that single example. Had the package **mi** been loaded, it would have automatically loaded **arm**, which in its turn would have loaded **MASS**. There could then be a confusion between the functions `select` in **MASS** and in **dplyr**. This kind of problem is becoming less frequent as packages 'import' other packages rather than 'depend' on them.

Graphics may require the correction of values in datasets, the calculation of new variables, or a restructuring of the data. R can handle as many datasets simultaneously as you wish, so it is essential to keep track of a dataset as changes are made. The best approach is to make all changes in R code and give the resulting dataset(s) new name(s). It is best to keep close tabs on which dataset or subset you are currently working with.

The examples in this book have been run using the package versions available at the end of 2014. Code chunks affected by subsequent changes in packages will be updated accordingly on the book's webpage, `rosuda.org/GDA`.

13.3   Graphics conventions in statistics

In drawing graphics it is important to be aware of the conventions and standards that are used. Not all will agree on these and additionally, as the classical phrase elegantly expresses it, the exception proves the rule. Nevertheless, most of the following are generally accepted and have been used without comment in the book. If something about a graphic surprises you, then perhaps you will find the explanation here.

1. In a scatterplot of causally related variables the dependent variable is drawn on the vertical axis and the explanatory variable on the horizontal axis.

2. Numbers increase to the right and up.

3. The x-axis crosses the y-axis at $y = 0$. When it doesn't, this should be made clear.

4. Scales are linear, and when they are not this should be emphasised.

5. Time is usually drawn on the horizontal axis, progressing from left to right.

6. Graphics are always drawn to show all the data. If some cases are out of range, this should be explicitly stated.

7. Aspect ratios (the relationship of the height of a graphic to its width) should be chosen to make the slope of lines of interest about $45°$, an idea first discussed thoroughly in [Cleveland et al., 1988].

8. Points usually represent individual cases and areas represent counts or weights.

9. Vertical bars represent frequencies of continuous variables when there are no gaps between them, and frequencies of categorical variables when there are.

10. Distinct colours are used to represent groups, while shading or continuous spectra are used to represent scales.

13.4   What is a graphic anyway?

In this book graphics are visual displays of data, summarising information about datasets. A graphic may show information about only one variable or about several and groups of simple graphics together are often more effective than a single complex graphic. Some graphics are easier to interpret and understand than others.

Most people are more ready to look at a graphic and try to understand it than to look at numbers or statistical output and try to understand that. However, it is probably also true that while people will tell you all too willingly that they do not understand statistics and cannot do simple maths, they will not tell you that they do not understand a graphic. Alan Fletcher makes a similar point in his stimulating book, "The Art of Looking Sideways" [Fletcher, 2001], where he remarks that people will say they have no ear for music, but will never admit to not being able to appreciate paintings. It is always a good idea to discuss graphics with others and to find out what they see and you do not—and vice versa.

There are many different graphic types used in data analysis: histograms, barcharts, dotplots, boxplots, scatterplots, and so on. Each type represents the data for one or more variables from a dataset in a graphical form. Sometimes each case is represented individually (as in a scatterplot), sometimes areas represent groups of cases (as in a barchart or a histogram).

When interpreting any graphic you have to make sure you understand how the plot has been defined and which data are shown. In general the plot types have standard definitions, which everyone follows, even if there are some exceptions. Several different definitions of boxplots are in use and however much we may feel that the classic Tukey definition is the one to use, we have to be aware that others may think otherwise.

It is tempting to think of graphics as just their graphics part, but there is much more to them than that. There are several components to a graphic above and beyond the graphical form itself:

Title There could be a title describing the graphic or directing your attention to some information in the graphic.

Caption The caption should explain what is shown, possibly also giving the data source. Captions should be detailed enough that the graphic can pretty well stand on its own. Longer is usually better than shorter. A picture may be worth a thousand words, but you need at least some words to describe and explain it.

Labels Well-labelled axes with understandable variable names (and, where appropriate, the units the data are measured in) make it easier to concentrate on the graphic, as there is no need to search for the information elsewhere.

Scales Uncluttered scales with nicely chosen numerical divisions that have meaning for the data help readers understand the orders of magnitude of the data on display.

Legend If there are different groups in the display with different colours or shapes, then a legend is valuable for defining them.

Annotations If a particular feature is to be highlighted, such as an outlier or a gap in the data, then an annotation can be added on the plot itself.

Accompanying text Graphics should always be discussed in the accompanying text or at the very least referred to. This provides an opportunity to comment on particular features in more detail and to add supplementary details. Ideally this text and the graphic should be close together. (Making sure this happens in a book is more difficult than you would think!)

Some of this applies more to presentation graphics than to exploratory graphics. Analysts exploring data will have much of the information at their fingertips and not need these details, they are looking for overall structure not exact numbers. Where the additional components do become relevant is when you discover an interesting piece of information and want to save it. The swift graphic that gave you the idea to-day may look a bit of a mystery in two weeks' time. Information that has been unearthed should be cleaned up and summarised and proper presentation graphics prepared, including most, if not all, of the components listed above. References to books with advice on drawing presentation graphics are given in §2.1, while [Unwin, 2008] is one of many articles that attempt to provide a summary of pertinent advice.

13.5   Options for all graphics

There is an astonishing number of different types of graphics you can draw for displaying data. Each also offers a broad range of individual options that can be chosen to modify its display even more. Amongst all this variety, there are options that apply equally to every graphic that should always be kept in mind.

Window size and shape

Size Plots look different at different sizes. Very small plots can be too cramped to be easy to see and extremely large plots can be difficult to digest. Often the same display leaves a different impression at different sizes. When comparing graphics of the same type it is imperative to plot them in the same size of display.

With presentation graphics there can be restrictions because of page size for printing or because of webpage structure if the graphics are displayed on the web. With exploratory graphics there is no reason for not experimenting with a range of sizes to ensure that you can find the best view for a plot. The sizes of displays in this book have been chosen to convey the information in the plots, while attempting to ensure that any text associated with the plots is on the same or the facing page. They have also been restricted by the size of the pages.

Aspect ratio Plots look more polished when they have a sensible aspect ratio, the ratio of the height to the width of a plot. They are also easier to interpret. Histograms which are broad and flat look squashed as if someone has trodden on them. If they are tall and thin, it is surprisingly difficult to judge the distributional shape. Both are actually unusual in practice, so try it yourself. If the histogram looks good to you in a default window, try it in a window which is twice as wide, but half as high, and in a window which is half as wide, but twice as high. One is too wide, one is too tall, and one may be just about right.

With boxplots it is another matter. One of their main advantages is their efficient use of space, a boxplot should be narrow and tall. Unfortunately many users of R seem to just use the default window shape, which results in wide, squat boxplots. Everyone is entitled to draw whatever plots they like in the privacy of their own analyses, but it is extraordinary how many of these boxes make it into print. Two wide boxplots side by side are not as easy to compare as two of appropriate width and they just do not look nice. A satisfactory solution is often to plot them horizontally rather than vertically.

With scatterplots it is usually best to use a square display, although sometimes other forms are better. If you crush one of the axes by reducing either the horizontal or vertical axis to almost nothing, you almost get a one-dimensional dotplot. Occasionally that can be useful (and quicker than drawing a new dotplot). A fine graphical example of the importance of getting the aspect ratio right can be found in a Doonesbury cartoon (http://doonesbury.slate.com/ strip/archive/2013/09/10).

Scales

Scaling Designing scales is difficult. Ideally you want sufficient discrimination, but not too much. You want to cover the full range of the data, and use rounded interpretable values. R uses the pretty algorithm designed for S [Becker et al., 1988] and in general it works impressively well and you can usually rely on the default choices. Several other algorithms have been suggested and can be found in the package **labeling**.

The way the pretty algorithm is implemented in base graphics has a curious side-effect, which you might call "walking the plank". To avoid drawing labels outside the range of the data (thereby increasing the size of the plot), while at the same time keeping 'pretty' values for the labels, the highest and lowest labels may not cover the full data range. This is a good idea, but what is not such a good idea is that the axes are only drawn as far as the final labels. Histogram bars to the far right can look as if they are falling off the edge or "walking the plank".

Common scaling for comparisons If two or more graphics are to be compared, it is essential to use common scaling, i.e., the same scaling for all. This includes the window size being the same if the plots are in different windows. Ensuring that windows are the same size can be achieved by using the default window size or by drawing new windows of a specific size using the height and width parameters. And, naturally, you should align the plots so that the comparisons you want to make are direct. Alignment is easier if carried out by R for you within a window, rather than trying to align separate plot windows by eye yourself.

Ensuring that scales are the same needs some preparatory work to check the value ranges for the axes, either by carrying out some initial test plotting or by doing the necessary calculations. For count and density scales, the vertical axes, the range needed can depend on binwidths (histograms) or bandwidths (density estimates). Precalculation of plot objects may be necessary. For categorical variables it means ensuring that the categories are in the same order in each plot and possibly adding one or more categories with zero counts to some of the variables to be compared. Test plotting is quick and can be informative in other ways, so that fits in with an exploratory approach. Calculations are useful for presentation graphics.

Scale limits The same plots look different with different scale limits. Sometimes there is good reason to include zero, even though it may be far from the data, sometimes not. Individual outliers can distort scales, leading to a lot of empty space and crowding most of the data into a small area.

Plots may be drawn in a box without a horizontal axis at zero. This can be misleading if readers think the lower boundary of the box is the zero line. It is advisable in such cases to add a faint grid line at zero.

Text

Labelling (abbreviations) Variable and category names can be informative and quite long. The need for Fortran's restriction to variable names that had at most six characters belongs to the distant past. This means that labelling for some plots suffers from overlapping text, as there is no way you can fit exorbitantly long names together for a barchart with many categories or for a parallel coordinate plot with many variables.

There are two solutions. The first is to create new variable names or new category names with appropriate abbreviations yourself. In both cases it is sensible to make a copy of the original dataset and make the changes in the copy. The second solution is to use R's `abbreviate` function. This works from left to right removing spaces and vowels until it reaches the allowed length, providing that no duplicates result. Again it is best to create a new dataset with the abbreviations. `abbreviate` is a considerable help, but does not work well on non-English names. Some manual adjusting afterwards may be worthwhile.

Font sizes The way the labellings of plots and other graphics texts look on screen and the way they look in print can be quite different. A lot depends on the original size of the plot and how it is changed for the printed page. This is something that is important for presentation graphics. It is at worst an irritant for exploratory graphics if the labels are too small to read or if they are so large that they overlap with each other or interfere with the rest of the graphic.

Annotations With presentation graphics it is sometimes a good idea to explicitly draw attention to a particular feature of a plot and use some form of annotation, adding some text and perhaps an arrow pointing to the feature. For best results it may be necessary to perfect the plot with specialist graphics software. Otherwise you can use some of the functions in base graphics, `annotate` in **ggplot2** or write your own function with **grid**. For exploratory graphics annotation is not necessary.

Colour and appearance

Colour Colour is a difficult topic for a wide variety of reasons. Fortunately there are several R packages that offer help. Two in particular are worth mentioning. **RColorBrewer** provides the colour palettes recommended by ColorBrewer [Brewer, 2013] for thematic maps. With **colorspace** you can map between different colour spaces including the recommended HCL [Zeileis et al., 2009]. There is also a good package vignette for **colorspace**. If colour is used, it is obviously best if possible colour blindness of viewers of the graphics is taken into account. Unfortunately default colour selections for some packages could be better, so you have to check.

One of the palettes used in this book is the colour blind one from the package **ggthemes** and another is from the **ggplot2** package based on palettes recommended by ColorBrewer [Brewer, 2013]. Palettes can be set at the start of a session. If you use R's default colour settings your displays will look different.

Colour is a powerful tool that should be used with care. Robert Simmon has written an attractive series of short web essays called "The Subtleties of Color" [Simmon, 2014], which offers some thought-provoking ideas on the topic. The best colour graphics are informative and tasteful. Anything less than the best can look kludgy and even be misleading (and you may well feel that this applies to some examples in the book).

'Themes' Not all plots look the same in this book, as they have been drawn with different R functions and packages. If you stick to one package, such as **ggplot2**, you can get a consistent look to your plots. You can specify 'themes' to ensure that the non-data parts of your plots are all in the same style and colour. I included the colourblind palette from **ggthemes** as part of the grey theme and also made the plot.background the same grey as the panel.background.

You have to specify the data parts of the plots separately, either with a general function at the beginning of a session or for each plot individually. For instance, I find the default black bars for barcharts and histograms too strong and set the default bar colour at the start of a session to achieve a standard grey colour, while ensuring that the borders of the bars can still be seen.

```
update_geom_defaults("bar",
          list(fill = I("grey70"), colour = I("grey40")))
```

13.6   Some R graphics advice and coding tips

A number of issues come up again and again in superficially different, but actually similar situations. The same principles generally apply, and this section comments on a few of them.

To get a new graphics window

(This assumes you are working with software which permits more than one graphics window. The popular RStudio interface basically only allows one.) The function dev.new() opens a default window whose size will depend on the R you are using. The units used, pixels, inches, or cms, can depend on your settings as well. When you want a bigger window than the default, or one with a particular aspect ratio, then specify the height and width:

```
dev.new(height=900, width=800)
```

Giving an exact size is useful when graphics suffer from overplotting in a default window, which can often happen with multivariate graphics, when you want to ensure that all windows in a group have the same non-default size, or when you are preparing graphics for a report or presentation.

Resizing windows

Sometimes the default window size is inappropriate for a graphic (e.g., text which overlaps in a small window will be fine in a larger one) or the aspect ratio makes the plot look wrong (e.g., overly wide boxplots). Rather than trying to predict the right height and width in advance it is simpler to just drag the window and choose the best size and aspect ratio by eye. You can then find out the size of the active window using `dev.size("px")` or `dev.size("in")`.

Default plots

One of R's strengths is the broad range of control options that most functions give users. It must be off-putting for beginners to look at the help for drawing a histogram, surely a simple task, and to wonder what on earth they have got themselves into, facing such a long list of options. Fortunately the complementary strength is that R's defaults are generally good. You can use `hist(x)` and get a reasonable plot and `plot(foo)` is useful for a wide range of objects `foo`, even if disappointing for a single numeric vector and not often helpful for a dataset. Not all of the defaults are as good as they could be, but that is hardly an issue when it is relatively easy to fix it yourself. You do not need to beware of defaults, but you do need to be wary of them.

Points in scatterplots or in other point plots

Scatterplot points are plotted as open circles by default in `base` graphics. For small datasets this is a matter of taste, for larger datasets it can look confusing (opinions may differ). Using one of the options `pch=16,19,20` in the plot function gives filled circles.

When scatterplots suffer from overplotting because of multiple cases with exactly the same values, various options have been suggested: special symbols to represent the numbers of cases as in sunflower plots (this is only reasonable for medium-sized datasets and the resulting graphics are still difficult to decode); jittering where small random values are added to the coordinates of the points to separate them (this can work, again only for datasets that are not really large); alpha-blending (where the shading of the points depends on how many there are); and hexagonal binning with shading. Zooming in to areas of particular interest can be valuable too. No approach is perfect and it is advisable to look at a range of plots and consider incorporating density estimates.

Printing graphics

The way graphics look on screen and the way they look printed from a saved file may be rather different, so you need to check. This can apply to colour, fonts, and spacing.

Multiple windows

Unlike some other systems, R requires you to open a new graphics window every time you draw a new plot, if you want to still keep existing plots. Its default behaviour is to draw a new graphic in the currently active graphics window. This can be irritating, and you have to get used to it. It is usually safest to always draw a new window, as you can lose track of which is the current active window. The code `dev.new()` has been taken for granted in much of the book.

This works differently for the RStudio interface, where you can use `x11()` instead, if you have access to an X Server, but that does take you out of RStudio. You then have some plots in X11 and some in RStudio. Within RStudio you do not overwrite old plots when you draw new ones, which is good, but it does not encourage you to see more than one plot at once, which is not so good. You can always flip through previously drawn graphics or, with sufficient advance planning, draw several in the same window.

Drawing several independent plots in one window

To draw a table of plots in base **graphics**, use `par(mfrow=c(n,m))`, which splits up the window into *n* rows and *m* columns. More complex layouts can be produced using `layout`. If you use a package based on **grid**, producing `grobs`, as **ggplot2** does, you can use the function `grid.arrange` from the **gridExtra** package to get a row and column layout. For more flexibility, you can combine `arrangeGrob` with `grid.arrange`, as was shown in Figure 12.2.

Figure 13.1 gives a simple example with two smaller plots above and one bigger plot below. First the plots are prepared as individual grobs and then the window to display the plots is set up. Observe how the grobs are nested and how the bigger plot is allocated more space.

If you want to use some of the useful graphics for categorical data in **vcd** you face the additional problem that while **vcd** is based on **grid** (like **ggplot2** and **lattice**), it uses an earlier version and does not produce the appropriate graphics objects (`grobs`). Even combining only **vcd** plots is hard work.

The two graphics systems in R, **graphics** and **grid**, do not mix well. There is a package **gridBase**, which in the words of its author, Paul Murrell, "can be used, in some situations, with a little care, to overcome this inherent incompatibility" [Murrell, 2011].

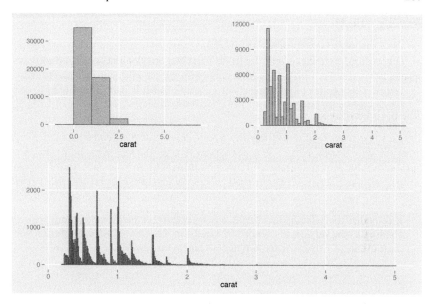

FIGURE 13.1: Three histograms of the variable `carat` from the *diamonds* dataset drawn together with different binwidths. The first plot suggests a coarse skew distribution. The second plot suggests there may be some favoured values. The third plot reveals the curious pattern of a number of skew distributions aligned together one after the other. The explanation is probably that rounded values of carat like 0.3, 0.5, 1, 1.2 are favoured with no diamonds of slightly smaller sizes being offered for sale and progressively fewer larger ones till the next rounded value is reached.

```
p0 <- ggplot(diamonds, aes(x=carat)) + ylab("")
p1 <- p0 + geom_bar(binwidth=1)
p2 <- p0 + geom_bar(binwidth=0.1)
p3 <- p0 + geom_bar(binwidth=0.01)
grid.arrange(arrangeGrob(p1, p2, ncol=2), p3, nrow=2)
```

Naming objects

- Give new objects new names
  If you drop, combine, or reorder levels in a categorical variable, it is best to create new variables with new names. Otherwise simple mistakes can lead to nasty errors. This applies to new objects of all kinds, but it is particularly easy to run into trouble with levels.

- Naming new variables
  Keep new variables associated with the data frame they are derived from.

```
iris <- within(iris, area <- Petal.Width*Petal.Length)
```

  will modify the existing data frame, *iris*, by adding an additional column, `area`, yielding a new version, also called *iris*, in the global environment. (Note that you must use the <- operator for assignment here; you cannot use =.) Whereas

```
area <- with(iris, area <- Petal.Width*Petal.Length)
```

  will generate a separate new variable, `area`, in the global environment using the columns of the data frame *iris*.

- Give plot objects names
  Functions like `hist(x)` plot the histogram in the current active window and `h1 <- hist(x); plot(h1)` does the same, but now you can access some of the components, e.g., `h1$counts`. Unfortunately, you cannot always access all you want. `hist` does not give you the binwidth directly, you have to calculate it from `with(h1, breaks[2]-breaks[1])`. This assumes that you know the bins are of equal size (highly recommended, only mathematical statisticians think of non-equal binsizes and they don't analyse real data very often) or that you have checked if `h1$equidist` is true. At a pinch you could use

```
binw <- with(h1, if(equidist) breaks[2]-breaks[1])
```

  Naming density estimates is particularly useful for getting at the bandwidth used.

Reordering categories for a barchart (and ordering in general)

The default ordering of categories is alphabetical. By redefining the levels attribute of a categorical variable you can change the order. To avoid confusion (and the possibility of mistakes) it is best to define a new variable. Here is a reordering of the car type variable by mean weight for the *Cars93* dataset in the **MASS** package:

```
data(Cars93, package="MASS")
Cars93 <- within(Cars93, TypeWt <- reorder(Type, Weight, mean))
```

You could achieve the same ordering by specifying the new variable's levels directly

```
Cars93 <- within(Cars93,
    Type1 <- factor(Type, levels=c("Small", "Sporty",
    "Compact", "Midsize", "Large", "Van")))
```

Compare the level orderings with

```
with(Cars93, table(TypeWt, Type1))
```

The value of choosing a good ordering can be seen in plots like Figure 13.2.

```
ggplot(Cars93, aes(TypeWt,100/MPG.city)) + geom_boxplot() +
ylab("Gallons per 100 miles") + xlab("Car type")
```

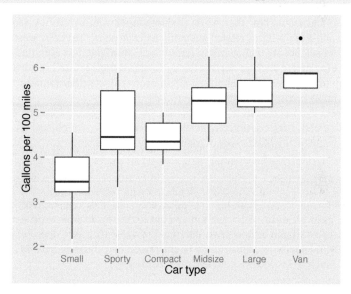

FIGURE 13.2: Fuel consumption by type of car ordered by average car weight. (It would look even better if the Sporty and Compact groups were switched.)

The replacement function `levels` can also be used to combine categories by repeating a value.

```
Cars93 <- within(Cars93, {
            Type3 <- Type
            levels(Type3) <- c("Small", "Large", "Midsize",
            "Small", "Sporty", "Large")
    })
```

Ordering can be useful for sorting cases (which can be important in overlaying one plot on another as in Figures 6.3 and 6.11), for sorting categories (for instance in the bottom two plots in Figure 4.1 or in the doubledecker plot in Figure 7.9), for sorting variables (as in Figures 6.11 and 6.15, cf. §6.7), and in drawing conditional plots like trellis displays.

The important functions are `order` for producing an ordered vector of case indices and `reorder` for ordering the levels of a factor. The functions `sort` and `rank` apply to single vectors. With `sort` you can order the vector and with `rank` get the ranks.

Reshaping datasets and graphics

Datasets frequently have to be restructured for particular graphics. As usual with R there are often several ways to achieve the same effects. In this book the packages **reshape2**, **tidyr**, and **dplyr** have been used, and they do require a certain amount of study to understand how they work. Both **reshape2** with its melting and casting functions and **tidyr** with its related gathering and spreading functions seem to need particular effort, but are well worth it. It must be pointed out, however, that it seems likely that the newer package **tidyr** will supersede **reshape2**. One big advantage of this group of packages is the ability to chain operations together (as was done, for instance, in the preparatory code for Figure 9.6).

Graphics themselves can be manipulated in several ways and facetting is a valuable general option for constructing small multiples of `ggplot` objects. The option `coord_flip()` is a handy way of transposing a ggplot display.

Missing values

Functions do not all deal with missing values the same way. For instance, compare what you get with `mean(X1)` to what you get for the mean with `summary(X1)`, when X1 is a variable with at least one missing value. If a plot does not work as you expect, then missing values could be the reason and you may have to set a plot function option or restructure your dataset accordingly.

Using the code and finding out about function options

This book is about the why of graphics not the how. There are a number of fine publications on how to draw graphics with R, so the emphasis here is different and there are occasionally fewer explanations of code details than some readers might like. If you need more information, R's help pages for functions are a good first place to look, although some are much better than others. Sometimes you find explanations and examples of a function's use covering the options available in comprehensive detail, sometimes not. Some packages have excellent vignettes, others not. If you need additional assistance, then the next step would be to type your query into your favourite search engine and that may well supply a solution to your problem.

13.7   Other graphics

This book has discussed analysing data graphically with some standard types of graphic. Perhaps it could have been called "The Seven Basic Plots", except that title has already been taken for discussing different kinds of plots by another book [Booker, 2004]. It is important to know the graphics you use well, so that you have experience in interpreting and understanding what they show. That is why the book has concentrated on only a few graphic types. The range of different patterns that can arise with them is surprising, and graphics can be very informative when you know what patterns to look for and what features might be revealed.

Of course, there are many other graphics, which you can consider and which might be especially useful for revealing particular information in certain datasets. Each of us may favour specific graphics forms over others but we should all be prepared to acknowledge that it is the information revealed, which is most important, not the graphic used.

The same principles apply to working with these additional graphics as have been discussed throughout the book. Amongst other things you should try many different variations of the plots, both in the options and the formatting, use more than one plot, ensure you have made informative comparisons—and not expect to find a single 'best' display. And if you present your plot to others bear in mind that they may not be as familiar with that type of graphic as you are.

Most of the graphics can be found in one or more packages in R in one form or another, and the Graphics Task View offers a convenient overview [Lewin-Koh, 2013]. You might draw: added variable plots, association plots, bagplots, Bangdiwala's agreement plots, barnest graphics, beanplots, beeswarm plots, bikini charts, biplots, Bland-Altman plots, bubble plots, bumps charts, centipede plots, coplots, corrgrams, cotab plots, coxcombs, dendrograms, Dickens' plots, donut plots, downfall plots, fan plots, forest plots, fourfold plots, funnel plots, hanging rootograms, hive plots, icicle plots, intersection plots, Kaplan-Meier plots, kite charts, L'Abbe plots, ladder plots, mountain plots, navel charts, Ord plots, pictograms, polarAnnulus plots, profile plots, pyramid plots, radar plots, ring diagrams, rose diagrams, Sankey diagrams, scree plots, sieve diagrams, silhouette plots, skyline plots, spaghetti plots, sparklines, speedometer charts, spider plots, spiecharts, stream graphs, sunflower plots, Taylor diagrams, ternary diagrams, tetris plots, thickening plots, tile plots, towel plots, trace plots, treemaps, triplets, violin plots, waffle plots, waterfall plots, ... to name but a few.

Apologies if your favourite plot is not included, no slight is intended. Apologies also that a couple of names that are not plots at all have also sneaked into the list, but now I can't remember which ones they are...

13.8   Large datasets

'Big Data' is a much discussed term nowadays and refers to datasets that are too large for common or garden software, whatever that implies at the time of the discussion and depending on what you want to do with the data. Some methods of analysis will work for pretty well any size of dataset and others are much more limited—just think of hierarchical clustering for example.

Even large datasets that can quite comfortably be analysed on a modern laptop bring special complications with them. From a statistical point of view, there are two problem areas. On the one hand, large numbers of variables lead to large numbers of tests, so that many are bound to be significant. On the other hand, large numbers of cases mean that even small effects will be declared significant in individual tests, although they may not be of practical importance.

There is another issue which lessens the relevance of these two, large datasets tend to be heterogeneous rather than homogeneous. Preliminary work may have to be done in selecting variables of interest and determining the subset of cases to use in an analysis before any testing is done, and the resulting dataset(s) may be much smaller. And if the dataset is still very large in the number of cases, it is worth taking advantage of the classical properties of statistical samples: there is no need to analyse the whole dataset, as its aggregate properties will be just as apparent in a random sample. To be on the safe side, we can either use fairly large samples or repeat an analysis with several different samples. This approach is fine for what you might call global properties of a dataset, but will not work for local ones such as unusual small groups.

There are some properties of large datasets that raise particular issues for graphics [Unwin et al., 2006]. There are almost certain to be some outliers, perhaps errors, on many variables and this distorts the scaling of initial plots, as we have seen for instance in Figures 3.11 and 9.8. Nominal variables may have so many categories that they need to be grouped to make displays informative [Eick and Karr, 2002]. In general there will be more of everything, more points, more bars, more lines, and plots need to be drawn larger. Zooming in becomes a useful tool and should be accompanied by an overall unzoomed view (a bird's-eye view) so that context is not lost.

Large datasets bring advantages too. You can examine ideas in much greater detail and check for more complex structures. If a hypothesis is developed on one part of the data, it may possibly be checked on another. If that cannot be done, it may still be feasible to explore the likely consequences of the idea and test it on other variables from the dataset not yet used in the analysis.

The final point to make about working with large datasets is also a graphical one: if there is a major effect, you should be able to draw a display that shows it.

13.9   Perfecting graphics

The emphasis in this book is on using graphics for data analysis, for exploring data rather than for formally presenting data. This means that it is important to be able to draw graphics quickly and flexibly, not having to worry about dotting the last i or crossing the last t of any displays. This demands reliable defaults for scales, legends, colour, aspect ratio, and all the other properties of graphics. In general R is good at that and even if it does not quite match up to your requirements in all respects, you can still prepare your own default versions of the graphics types you use most often to use instead.

If you have to present graphics in a publication, you may have to be much more concerned with details: the fonts used, the precise colours, the placing of legends, supplementary annotations and so on. It is possible to specify everything you need in R, it is just that sometimes it takes rather a lot of code and some juggling with the options. There is every reason to prepare as perfect a graphic as you can, once you know which graphic you want and provided you have the time. Getting a display just right for presentation is a different mode of working to carrying out analyses of data and it is probably best to keep the two activities separate. Ideally neither design issues nor computing problems should interfere with the analysis process.

For exploratory work the aim should be to produce flexible and informative graphics, looking at many useful, if unpolished, graphics, to get an overall picture, rather than concentrating on one elegant graphic that may not reveal the full story. To paraphrase John Tukey: Better to draw several approximate graphics saying something about the right question than to draw one precise graphic relating to the wrong question.

14

Summary

The single biggest problem in communication is the illusion that it has taken place.

George Bernard Shaw

14.1   Data analysis and graphics

Good graphics are informative, effective and flexible. Graphics can be attractive and encourage discussion, they can be more insightful and convincing than text. They are fairly easy to explain to others, often easier to explain than the results of statistical analyses. There is every reason to use them more in data analysis.

With the benefit of hindsight, Graphical Data Analysis is simple, you just show the information contained in the data. In practice that is less straightforward than it sounds, you first have to find out what information is there. The approach described in this book recommends using a collection of common graphic displays to help you uncover information. This is not about looking for a single optimal display and showing how to draw the plot as attractively and precisely as possible. Rather it is about drawing many displays and aiming to find an optimal set. Here 'optimal' means that taken as a group the set of graphics show a lot—if not all—of the information that is in the data.

Different types of graphics highlight different features in datasets. It is worth drawing a few to check what you can see. A group of graphics of the same type, but with different formatting and scaling, may also pick out a range of different aspects of data. It is useful to experiment with different options: growing or shrinking plots, changing their aspect ratios, reordering categories, and trying out different binwidths or bandwidths. Of course, it is all too easy to get lost in large numbers of variables and piles of output. Successful analyses, particularly of big datasets, depend on good organisation. Ideally you need a manager for your analyses like the housekeeper played by Helen Mirren in the film "Gosford Park", who knew in advance what the guests needed, before they knew it themselves.

14.2   Key features of GDA

GDA entails drawing and interpreting graphics, the drawing alone is not enough. "Every Picture Tells a Story" (Rod Stewart) and it is an essential part of GDA to look for the story in every picture. A graphic with no message is no use.

GDA requires a strategic approach. It is about using many graphics at once and being prepared to consider several different lines of thought in parallel. Who can tell which of the features identified in graphics may turn out to be the most interesting? It is important to have overall goals and simultaneously to be flexible in pursuing the general goal of discovering new information.

Graphics may look different because of how they are drawn—the formatting, scaling, and colouring. Using many graphics avoids possibly misleading effects in individual displays due to this. On the other hand the same graphic may be interpreted differently because of the size of the underlying dataset or the subject matter context. Any such difference is intrinsic and should be emphasised not avoided.

GDA is about the generation of ideas, not just data description, and not the testing of ideas, the contrast Tukey was referring to in writing of detectives and judges. GDA should be used in close association with statistical modelling, as the two approaches complement each other well. Testing tells you if a null hypothesis should be rejected but not necessarily why. Graphics show you what might be 'wrong', but gives no guidance on how strong the evidence is. Statistical tests are useful for checking whether an interesting graphical feature is significant. Graphics are useful for asking whether a statistically significant result is really of interest

You can't prove anything with statistics, but you can disprove some things and the same goes for graphics. Drawing many graphics quickly and informatively for exploratory purposes is also different from drawing a few graphics attractively and precisely for presentation purposes. Tukey might have suggested the contrast of detectives and designers.

14.3   Strengths and weaknesses of GDA

Every approach has its pluses and minuses. In the case of GDA there can be downsides due to badly chosen graphics, overloaded graphics, poorly organised groups of graphics, over-interpretation of graphics, and apophenia (seeing patterns where there aren't any). GDA is, of course, also affected if data are inadequate, a problem for all statistical and data analyses.

The strengths of GDA lie in its flexibility and ease of communication. It is also rather robust. If you make an error or draw the wrong plot, you can readily see what has happened and fix the problem. GDA is good for data cleaning, exploring data structure, detecting outliers and unusual groups, identifying trends and clusters, spotting local patterns, evaluating modelling output, and presenting results. It is about generating ideas, while statistics is more about evaluating ideas.

14.4 Recommendations for GDA

A number of recommendations recur in this book:

- Use graphics to discover information that is difficult to investigate statistically.

- Draw many graphics and vary the graphics options.

- Gain experience in interpreting graphics and use the graphics types you know well.

- Consider reformatting datasets before drawing graphics.

- Make appropriate comparisons and choose comparable scales.

- Check any graphical result you find with statistical models, where possible. Statistics and graphics complement one another.

- Always remember how important context is in interpreting results.

The same statistics and graphics might be interpreted quite differently for different applications. That is why this book has concentrated on using 'real' datasets. Finally, interactive graphics have occasionally been mentioned in this book as a tool with potential for GDA and that is an important topic for the future.

References

[Aggarwal, 2013] Aggarwal, C. (2013). *Outlier Analysis*. Springer, New York.

[Agresti, 2007] Agresti, A. (2007). *Categorical Data Analysis*. Wiley, New York, 2nd edition.

[Albert, 1994] Albert, J. (1994). Exploring baseball hitting data: What about those breakdown statistics? *Journal of the American Statistical Association*, 89(427):1066–1074.

[ASA, 2014] ASA (2014). Publications of the American Statistical Association. `http://amstat.tandfonline.com`. (accessed 18.10.2014).

[Bache and Lichman, 2013] Bache, K. and Lichman, M. (2013). UCI machine learning repository. `http://archive.ics.uci.edu/ml`. (accessed 12.11.2014).

[Barnett, 1994] Barnett, A. (1994). How numbers can trick you. *Technology Review*, 97(7):38–45.

[Barnett and Lewis, 1994] Barnett, V. and Lewis, T. (1994). *Outliers in Statistical Data*. Wiley, Chichester, 3rd edition.

[Becker et al., 1988] Becker, R., Chambers, J., and Wilks, A. (1988). *The New S Language*. Wadsworth & Brooks/Cole, Pacific Grove, California.

[Becker et al., 1996] Becker, R. A., Cleveland, W. S., and Shyu, M. J. (1996). The visual design and control of trellis display. *Journal of Computational and Graphical Statistics*, 5(2):123–155.

[Bederson and Shneiderman, 2003] Bederson, B. and Shneiderman, B. (2003). *The Craft of Information Visualization: Readings and Reflections*. Morgan Kaufmann, San Francisco, California.

[Ben-Gal, 2005] Ben-Gal, I. (2005). Outlier detection. In Maimon, O. and Rokach, L., editors, *Data Mining and Knowledge Discovery Handbook*, pages 131–146. Springer, New York.

[Benford, 1938] Benford, F. (1938). The law of anomalous numbers. *Proceedings of the American Philosophical Society*, 78(4):551–572.

[Benjamini, 2010] Benjamini, Y. (2010). Discovering the false discovery rate. *Journal of the Royal Statistical Society: Series B (Statistical Methodology)*, 72(4):405–416.

[Bertin, 2010] Bertin, J. (2010). *Semiology of Graphics*. Esri Press, Redlands, California.

[Bissantz, 2009] Bissantz, N. (2009). Me, myself, and BI. http://blog.bissantz.com/vis-a-vis. (accessed 14.12.2013).

[Booker, 2004] Booker, C. (2004). *The Seven Basic Plots*. Continuum, London.

[Brain, 2004] Brain, M. (2004). How alligators work. http://animals.howstuffworks.com/reptiles/alligator3.htm. (accessed 12.10.2014).

[Brant and Nguyen, 2008] Brant, S. and Nguyen, G. (2008). Is there a gender difference in the prevalence of Crohn's disease or ulcerative colitis? *Inflammatory Bowel Diseases*, 14(2).

[Bretz et al., 2010] Bretz, F., Hothorn, T., and Westfall, P. (2010). *Multiple Comparisons Using R*. Chapman & Hall/CRC Press, Boca Raton, Florida, USA.

[Brewer, 2013] Brewer, C. (2013). ColorBrewer. http://www.colorbrewer.org. (accessed 12.11.2014).

[Chang, 2012] Chang, W. (2012). *R Graphics Cookbook*. O'Reilly Media, Beijing.

[Chen et al., 2008] Chen, C., Haerdle, W., and Unwin, A. R. (2008). *Handbook of Data Visualization*. Springer, Berlin.

[Clemons and Pagano, 1999] Clemons, T. and Pagano, M. (1999). Are babies normal? *The American Statistician*, 53(4):298–302.

[Cleveland, 1994] Cleveland, W. (1994). *The Elements of Graphing Data*. Hobart Press, Summit, New Jersey, revised edition.

[Cleveland et al., 1982] Cleveland, W., Diaconis, P., and McGill, R. (1982). Variables on scatterplots look more highly correlated when the scales are increased. *Science*, 216:1138–41.

[Cleveland et al., 1988] Cleveland, W., McGill, M. E., and McGill, R. (1988). The shape parameter of a two-variable graph. *Journal of the American Statistical Association*, 83(402):289–300.

[Cleveland and McGill, 1987] Cleveland, W. and McGill, R. (1987). Graphical perception: The visual decoding of quantitative information on graphical displays of data. *Journal of the Royal Statistical Society A*, 150(3):192–229.

[Cleveland, 1993] Cleveland, W. S. (1993). *Visualizing Data*. Hobart Press, Summit, New Jersey.

[Coen et al., 1969] Coen, P., Gomme, D., and Kendall, M. (1969). Lagged relationships in economic forecasting. *Journal of the Royal Statistical Society A*, 132(2):133–163.

[Cook and Swayne, 2007] Cook, D. and Swayne, D. (2007). *Interactive and Dynamic Graphics for Data Analysis*. Springer, New York.

[Davison, 2008] Davison, A. (2008). *Statistical Models*. Cambridge University Press, Cambridge.

[De Veaux et al., 2011] De Veaux, R., Velleman, P., and Bock, D. (2011). *Stats: Data and Models*. Pearson, Boston.

[Dodge, 1996] Dodge, Y. (1996). The guinea pig of multiple regression. In Rieder, H., editor, *Robust Statistics, Data Analysis, and Computer Intensive Methods*, pages 91–117. Springer, New York.

[DuToit et al., 1986] DuToit, S., Steyn, A., and Stumpf, R. (1986). *Graphical Exploratory Data Analysis*. Springer, New York.

[Edwards et al., 1963] Edwards, W., Lindman, H., and Savage, L. (1963). Bayesian statistical inference for psychological research. *Psychological Review*, 70(3):193–242.

[Eick and Karr, 2002] Eick, S. and Karr, A. (2002). Visual scalability. *Journal of Computational and Graphical Statistics*, 11(1):22–43.

[Fahrmeir et al., 2013] Fahrmeir, L., Kneib, T., and Lang, S. (2013). *Regression*. Springer, Berlin.

[Few, 2012] Few, S. (2012). *Show Me the Numbers: Designing Tables and Graphs to Enlighten*. Analytics Press, Burlingame, California, 2nd edition.

[Fisher, 1925] Fisher, R. (1925). *Statistical Methods for Research Workers*. Oliver and Boyd, Edinburgh.

[Fisher, 1936] Fisher, R. (1936). The use of multiple measurements in taxonomic problems. *Annals of Eugenics*, 7(2):179–188.

[FiveThirtyEight, 2014] FiveThirtyEight (2014). Why Pollsters Think They Underestimated 'No' in Scotland. http://fivethirtyeight.com/datalab/why-pollsters-think-they-underestimated-no-in-scotland/. (accessed 30.11.2014).

[Fletcher, 2001] Fletcher, A. (2001). *The Art of Looking Sideways*. Phaidon, London.

[Flury and Riedwyl, 1988] Flury, B. and Riedwyl, H. (1988). *Multivariate Statistics A Practical Approach*. Chapman & Hall, London.

[Forina et al., 1983] Forina, M., Armanino, C., Lanteri, S., and Tiscornia, E. (1983). Classification of olive oils from their fatty acid composition. In Martens, H. and Russwurm, H. J., editors, *Food Research and Data Analysis*, pages 189–214. Applied Science Publishers, London.

[Forina et al., 1988] Forina, M., Leardi, R., Armanino, C., and Lanteri, S. (1988). Parvus - an extendible package for data exploration, classification and correlation.

[Freedman et al., 2007] Freedman, D., Pisani, R., and Purves, R. (2007). *Statistics.* W.W. Norton, New York, 4th edition.

[Frey et al., 2011] Frey, B., Savage, D., and Torgler, B. (2011). Behavior under extreme conditions: The Titanic disaster. *Journal of Economic Perspectives*, 25:209–222.

[Friendly, 1994] Friendly, M. (1994). Mosaic displays for multi-way contingency tables. *Journal of the American Statistical Association*, 89(1):190–200.

[Friendly, 2000] Friendly, M. (2000). *Visualizing Categorical Data.* SAS, Cary, North Carolina.

[Friendly, 2011] Friendly, M. (2011). Gallery of data visualization. http://www.datavis.ca/gallery/. (accessed 12.11.2014).

[Fung, 2011] Fung, K. (2011). Junk charts. http://junkcharts.typepad.com/junk_charts/. (accessed 12.11.2014).

[Gannett, 1898] Gannett, H. (1898). *Statistical Atlas of the United States.* Washington, D.C.: Government Print Office. (available at http://www.census.gov/history/www/programs/geography/statistical_atlases.html, accessed 20.10.2014).

[Gastwirth, 2006] Gastwirth, J. (2006). A 60 million dollar statistical issue arising in the interpretation and calculation of a measure of relative disparity: Zuni Public School District 89 v. U.S. Department of Education. *Law Probability & Risk*, 5(1):33–61.

[Gastwirth, 2008] Gastwirth, J. (2008). The U.S. Supreme Court finds a statute's description of a simple statistical measure of relative disparity 'ambiguous' allowing the Secretary of Education to interpret the formula: Zuni Public School District 89 v. U.S. Department of Education II. *Law Probability & Risk*, 7(3):225–248.

[Geary, 1947] Geary, R. C. (1947). Testing for normality. *Biometrika*, 34:209–242.

[Gelman, 2011] Gelman, A. (2011). Statistical modeling, causal inference, and social science. http://www.stat.columbia.edu/~gelman/blog/. (accessed 12.11.2014).

[Gelman et al., 2013] Gelman, A., Carlin, J., Stern, H., Dunson, D., Vehtari, A., and Rubin, D. (2013). *Bayesian Data Analysis.* Chapman & Hall/CRC, Boca Raton, Florida, 3rd edition.

[Gelman and Hill, 2006] Gelman, A. and Hill, J. (2006). *Data Analysis Using Regression and Multilevel/Hierarchical Models*. Cambridge University Press, Cambridge.

[Gilby, 1911] Gilby, W. H. (1911). On the significance of the teacher's appreciation of general intelligence. *Biometrika*, 8:94–108.

[Good and Gaskins, 1980] Good, I. J. and Gaskins, R. A. (1980). Density estimation and bump-hunting by the penalized likelihood method exemplified by scattering and meteorite data. *Journal of the American Statistical Association*, 75(369):42–56.

[Google, 2010] Google (2010). Ngram viewer. `https://books.google.com/ngrams`. (accessed 12.11.2014).

[Gregg, 1994] Gregg, P. (1994). Out for the count: A social scientist's analysis of unemployment statistics in the UK. *Journal of the Royal Statistical Society A*, 157(2):253–270.

[Guardian, 2013] Guardian (2013). University guide 2013. `http://www.guardian.co.uk/news/datablog/2012/may/22/university-guide-2013-guardian-data`. (accessed 12.11.2014).

[Hanley, 2004] Hanley, J. A. (2004). "Transmuting" women into men: Galton's family data on human stature. *The American Statistician*, 58(3):237–243.

[Hartigan, 1975] Hartigan, J. A. (1975). *Clustering Algorithms*. John Wiley & Sons, Inc., New York.

[Hartigan and Hartigan, 1985] Hartigan, J. A. and Hartigan, P. M. (1985). The dip test of unimodality. *The Annals of Statistics*, 13(1):70–84.

[Hartigan and Kleiner, 1981] Hartigan, J. A. and Kleiner, B. (1981). Mosaics for contingency tables. In *Computer Science and Statistics: Proceedings of the 13th Symposium on the Interface*, pages 268–273, New York. Springer.

[Hastie et al., 2001] Hastie, T., Tibshirani, R., and Friedman, J. (2001). *The Elements of Statistical Learning*. Springer, New York, 2nd edition.

[Hummel, 1996] Hummel, J. (1996). Linked bar charts: Analysing categorical data graphically. *Computational Statistics*, 11:23–33.

[Hurley and Oldford, 2010] Hurley, C. and Oldford, W. (2010). Pairwise display of high-dimensional information via Eulerian tours and Hamiltonian decompositions. *Journal of Computational and Graphical Statistics*, 19(4):861–886.

[Hyndman, 2013] Hyndman, R. J. (2013). Time series task view. `http://cran.r-project.org/web/views/TimeSeries.html`. (accessed 12.11.2014).

[IAAF, 2001] IAAF (2001). *Scoring Tables for Combined Events*. IAAF, Monaco.

[IBM, 2007] IBM (2007). Many eyes. `http://many-eyes.com/`. (accessed 12.11.2014).

[Inselberg, 2009] Inselberg, A. (2009). *Parallel Coordinates: Visual Multidimensional Geometry and Its Applications*. Springer, New York.

[Izenman, 2008] Izenman, A. (2008). *Modern Multivariate Statistical Techniques*. Springer, New York.

[Izenman and Sommer, 1988] Izenman, A. and Sommer, C. (1988). Philatelic mixtures and multimodal densities. *Journal of the American Statistical Association*, 83(404):941–953.

[James, 2014] James, S. (2014). US and UK inflation. `http://www.sscnet.ucla.edu/polisci/faculty/james/download/inflation.xls`. (accessed 12.11.2014).

[Johnson and Shneiderman, 1991] Johnson, B. and Shneiderman, B. (1991). Treemaps: a space-filling approach to the visualization of hierarchical information structures. In *Proceedings of the IEEE Conference on Visualization '91*, pages 284–291.

[Kleiber and Zeileis, 2008] Kleiber, C. and Zeileis, A. (2008). *Applied Econometrics with R*. Springer, New York.

[Kosara, 2011] Kosara, R. (2011). eagereyes. `http://eagereyes.org/`. (accessed 12.11.2014).

[Kosara et al., 2006] Kosara, R., Bendix, F., and Hauser, H. (2006). Parallel sets: Visual analysis of categorical data. *Transactions on Visualization and Computer Graphics*, 12(4):558–568.

[Krause and McConnell, 2012] Krause, A. and McConnell, M. (2012). *A Picture is Worth a Thousand Tables: Graphics in Life Sciences*. Springer, New York.

[Lewin-Koh, 2013] Lewin-Koh, N. (2013). Graphics task view. `http://cran.r-project.org/web/views/Graphics.html`. (accessed 12.11.2014).

[Lock, 1993] Lock, R. (1993). 1993 new car data. *Journal of Statistics Education*, 1(1).

[Lumley, 2010] Lumley, T. (2010). *Complex Surveys: A Guide to Analysis Using R*. Wiley, Hoboken, New Jersey.

[Maindonald and Braun, 2010] Maindonald, J. and Braun, R. (2010). *Data Analysis and Graphics Using R*. Cambridge University Press, Cambridge.

[Malik, 2010] Malik, W. (2010). *Data Cleaning of Large Datasets: New Methods and Techniques*. Verlag Dr. Hut, Munich.

[Mardia et al., 1979] Mardia, K. V., Kent, J. T., and Bibby, J. M. (1979). *Multivariate Analysis*. Academic Press, London.

[Meyer et al., 2006] Meyer, D., Zeileis, A., and Hornik, K. (2006). The strucplot framework: Visualizing multi-way contingency tables with vcd. *Journal of Statistical Software*, 17(3):1–48.

[Mihos, 2005] Mihos, C. (2005). Astronomy 221. `http://burro.astr.cwru.edu/Academics/Astr221/HW/HW5/HW5.html`. (accessed 17.11.2014).

[Mittal, 2011] Mittal, H. (2011). *R Graph Cookbook*. PACKT PUBLISHING, Birmingham.

[Murrell, 2005] Murrell, P. (2005). *R Graphics*. Chapman & Hall, London.

[Murrell, 2009] Murrell, P. (2009). *Data Technologies*. Chapman & Hall, London.

[Murrell, 2011] Murrell, P. (2011). *R Graphics*. Chapman & Hall, London, 2nd edition.

[New York Times, 2011] New York Times (2011). New York Times Infographics. `http://www.smallmeans.com/new-york-times-infographics/`. (accessed 17.11.2014).

[Newcomb, 1881] Newcomb, S. (1881). Note on the frequency of use of the different digits in natural numbers. *American Journal of Mathematics*, 4(1):39–40.

[Norman, 1988] Norman, D. A. (1988). *The Design of Everyday Things*. MIT Press, Cambridge, Mass.

[OCSI, 2009] OCSI (2009). Improving data visualisation for the public sector. `http://www.improving-visualisation.org/`. (accessed 12.11.2014).

[Olsen, 1998] Olsen, C. (1998). A comparison of parametric and semiparametric estimates of the effect of spousal health insurance coverage on weekly hours worked by wives. *Journal of Applied Econometrics*, 13(5):543–565.

[Ord and Fildes, 2013] Ord, K. and Fildes, R. (2013). *Principles of Business Forecasting*. Cengage Learning, Mason, Ohio.

[Pearson and Lee, 1903] Pearson, K. and Lee, A. (1903). On the laws of inheritance in man: I. Inheritance of physical characters. *Biometrika*, 2(4):357–462.

[Pilhoefer et al., 2012] Pilhoefer, A., Gribov, A., and Unwin, A. (2012). Comparing clusterings using Bertin's idea. *IEEE Transactions on Visualization and Computer Graphics*, 18(12):2506–2515.

[Pilhoefer and Unwin, 2013] Pilhoefer, A. and Unwin, A. (2013). New approaches in visualization of categorical data: R-package extracat. *Journal of Statistical Software*, 53(7):1–25.

[Playfair, 2005] Playfair, W. (2005). *Playfair's Commercial and Political Atlas and Statistical Breviary*. Cambridge University Press, Cambridge.

[Rahlf, 2014] Rahlf, T. (2014). *Datendesign mit R — 100 Visualisierungsbeispiele*. Open Source Press, Munich.

[rlearnr, 2009] rlearnr (2009). Learning R. http://learnr.wordpress.com/. (accessed 12.11.2014).

[Robbins, 2005] Robbins, N. (2005). *Creating More Effective Graphs*. Wiley-Blackwell, Hoboken, New Jersey.

[Rogers and Hsu, 2001] Rogers, J. and Hsu, J. (2001). Multiple comparisons of biodiversity. *Biometrical Journal*, 43(5):617–625.

[Rosling, 2013] Rosling, H. (2013). Gapminder. http://www.gapminder.org. (accessed 12.11.2014).

[Rubin, 1987] Rubin, D. (1987). *Multiple Imputation for Nonresponse in Surveys*. Wiley, New York.

[Salmistu, 2013] Salmistu, J. (2013). Decathlon2000. http://www.decathlon2000.com/eng/. (accessed 12.11.2014).

[Sarkar, 2008] Sarkar, D. (2008). *Lattice: Multivariate Data Visualization with R*. useR. Springer, New York.

[Schilling et al., 2002] Schilling, M. F., Watkins, A. E., and Watkins, W. (2002). Is human height bimodal? *The American Statistician*, 56(3):223–229.

[Scott, 1992] Scott, D. (1992). *Multivariate Density Estimation — Theory, Practice, and Visualization*. Wiley, New York.

[Sheehan et al., 1991] Sheehan, P., Fastovsky, D., Hoffmann, R. G., Berghaus, C., and Gabrielt, D. (1991). Sudden extinction of the dinosaurs: Latest Cretaceous, Upper Great Plains, U.S.A. *Science*, 254(5033):835–839.

[Simmon, 2014] Simmon, R. (2014). Subtleties of color. http://earthobservatory.nasa.gov/blogs/elegantfigures/2013/08/05/subtleties-of-color-part-1-of-6/. (accessed 13.10.2014).

[Snee, 1974] Snee, R. (1974). Graphical display of two-way contingency tables. *The American Statistician*, 28(1):9–12.

[Spence, 2007] Spence, R. (2007). *Information Visualization: Design for Interaction*. Pearson, Harlow, England.

[Sunday Independent, 2013] Sunday Independent (2013). FG and FF neck and neck in poll. `http://www.independent.ie/irish-news/fg-and-ff-neck-and-neck-in-poll-29507501.html`. (accessed 17.11.2014).

[*Guardian*, 2011] *Guardian* (2011). Datastore. `http://www.guardian.co.uk/data`. (accessed 12.11.2014).

[Theus, 2013] Theus, M. (2013). Statistical graphics and more. `http://www.theusRus.de/blog/`. (accessed 12.11.2014).

[Theus and Urbanek, 2007] Theus, M. and Urbanek, S. (2007). *Interactive Graphics for Data Analysis*. CRC Press, London.

[Thode Jr., 2002] Thode Jr., H. (2002). *Testing for Normality*. Marcel Dekker, New York.

[TimelyPortfolio, 2013] TimelyPortfolio (2013). R financial time series plotting. `http://timelyportfolio.github.io/rCharts_time_series/history.html`. (accessed 12.11.2014).

[Tsay, 2014] Tsay, R. (2014). *Multivariate Time Series Analysis: with R and Financial Applications*. Wiley, Hoboken, New Jersey.

[Tufte, 1990] Tufte, E. (1990). *Envisioning Information*. Graphic Press, Cheshire, Connecticut.

[Tufte, 2001] Tufte, E. (2001). *The Visual Display of Quantitative Information*. Graphic Press, Cheshire, Connecticut, 2nd edition.

[Tufte, 2013] Tufte, E. (2013). ET NOTEBOOKS. `http://www.edwardtufte.com/bboard/`. (accessed 12.11.2014).

[Tukey, 1993] Tukey, J. (1993). Graphic comparisons of several linked aspects: Alternatives and suggested principles. *Journal of Computational and Graphical Statistics*, 2(1):1–33.

[Tutz, 2012] Tutz, G. (2012). *Regression for Categorical Data*. Cambridge University Press, Cambridge.

[Unwin et al., 2013] Unwin, A., Hofmann, H., and Cook, D. (2013). Let graphics tell the story - datasets in R. *R Journal*, 5(1):117–129.

[Unwin, 2008] Unwin, A. R. (2008). Good graphics? In Chen, C., Haerdle, W., and Unwin, A. R., editors, *Handbook of Data Visualization*, pages 57–78. Springer, New York.

[Unwin et al., 2006] Unwin, A. R., Theus, M., and Hofmann, H. (2006). *Graphics of Large Datasets*. Springer, New York.

[Unwin et al., 2003] Unwin, A. R., Volinsky, C., and Winkler, S. (2003). Parallel coordinates for exploratory modelling analysis. *Computational Statistics & Data Analysis*, 43(4):553–564.

[Urbanek et al., 2014] Urbanek, S., Woodhull, G., et al. (2014). Rcloud. `http://stats.research.att.com/RCloud/`. (accessed 12.11.2014).

[van Belle et al., 2004] van Belle, G., Fisher, L., Heagerty, P., and Lumley, T. (2004). *Biostatistics: A Methodology For the Health Sciences*. Wiley, New York, 2nd edition.

[Vande Moere, 2011] Vande Moere, A. (2011). information aesthetics. `http://infosthetics.com/`. (accessed 12.11.2014).

[Venables and Ripley, 2002] Venables, W. N. and Ripley, B. D. (2002). *Modern Applied Statistics with S*. Springer, New York, 4th edition.

[Viz, 2011] Viz (2011). Visualizing.org. `http://www.visualizing.org/`. (accessed 12.11.2014).

[von Bortkiewicz, 1898] von Bortkiewicz, L. (1898). *Das Gesetz der kleinen Zahlen*. Teubner, Leipzig.

[Wainer, 1997] Wainer, H. (1997). *Visual Revelations*. Springer, New York.

[Wainer, 2004] Wainer, H. (2004). *Graphic Discovery: A Trout in the Milk and Other Visual Adventures*. Princeton University Press, Princeton, New Jersey.

[Wainer, 2009] Wainer, H. (2009). *Picturing the Uncertain World: How to Understand, Communicate, and Control Uncertainty through Graphical Display*. Princeton University Press, Princeton, New Jersey.

[Wand, 1997] Wand, M. (1997). Data-based choice of histogram bin width. *The American Statistician*, 51(1):59–64.

[Wand and Jones, 1995] Wand, M. and Jones, M. (1995). *Kernel Smoothing*. Chapman & Hall, London.

[Ware, 2008] Ware, C. (2008). *Visual Thinking: for Design*. Morgan Kaufmann, San Francisco, California.

[Wattenberg, 2005] Wattenberg, M. (2005). Name voyager. `http://www.babynamewizard.com/voyager`. (accessed 11.11.2014).

[Wegman, 1990] Wegman, E. (1990). Hyperdimensional data analysis using parallel coordinates. *Journal of the American Statistical Association*, 85:664–675.

[Wickham, 2009] Wickham, H. (2009). *ggplot2: Elegant graphics for data analysis*. useR. Springer, New York.

[Wikipedia, 2013] Wikipedia (2013). Old Faithful. `http://en.wikipedia.org/wiki/Old_Faithful`. (accessed 17.11.2014).

[Wikipedia, 2014] Wikipedia (2014). Charge of the Light Brigade. `http://en.wikipedia.org/wiki/Charge_of_the_Light_Brigade`. (accessed 21.3.2014).

[Wilkinson, 2005] Wilkinson, L. (2005). *The Grammar of Graphics*. Springer, New York, 2nd edition.

[Wills, 2012] Wills, G. (2012). *Visualizing Time*. Springer, New York.

[Worcester, 1996] Worcester, R. (1996). Political polling: 95% expertise and 5% luck. *Journal of the Royal Statistical Society. Series A (Statistics in Society)*, 159(1):5–20.

[Wright, 2013] Wright, K. (2013). Revisiting Immer's barley data. *The American Statistician*, 67(3):129–133.

[Yau, 2011] Yau, N. (2011). FlowingData. `http://flowingdata.com/`. (accessed 17.11.2014).

[Yule, 1926] Yule, G. (1926). Why do we sometimes get nonsense-correlations between time-series?—A study in sampling and the nature of time-series. *Journal of the Royal Statistical Society*, 89(1):1–63.

[Zeileis et al., 2009] Zeileis, A., Hornik, K., and Murrell, P. (2009). Escaping rgb-land: Selecting colors for statistical graphics. *Computational Statistics & Data Analysis*, 53:3259–3270.

[Zheng and Gastwirth, 2010] Zheng, T. and Gastwirth, J. (2010). On bootstrap tests of symmetry about an unknown median. *Journal of Data Science*, 8:397–412.

General index

Datasets index